MATHEMATICS AND THE UNIVERSE

An Interpretation Based on the Theory of Relativity

E. T. Lawrence

VANTAGE PRESS
New York Washington Atlanta Hollywood

FIRST EDITION

Published by Vantage Press, Inc.
516 West 34th Street, New York, New York 10001

Printed in the United States of America
Standard Book Number 533-02537-0

PREFACE

At one time, matter was considered to be made of indestructible particles called atoms. Recently, the Theory of Relativity, as proposed by Einstein, has provided the tools to break up the atom. It is the premise of this book that the Theory of Relativity can provide the tools for exceptions to other concepts that, until now, were regarded as absolute.

One of these absolutes is the "equals" function. This function, which is basic to mathematics, may be expressed by the logic that if *A* equals *B*, and *A* equals *C*, then *B* equals *C*. A well-known thought experiment in Relativity Theory can be interpreted to mean that, in the real universe, if *A* equals *B* and *A* equals *C*, *B* does not necessarily equal *C*.

This does not mean the equals function is invalid, but it does mean that it is not absolute. In addition, if there is not necessarily a strict adherence to the equals function in the universe, then the notion of creation and destruction can be credibly hypothesized. Creation and destruction can be expressed mathematically as the breaking of the Law of Conservation of Energy and Law of Conservation of Momentum, until now also regarded as two of the absolutes in the universe. The final portion of this book is devoted to showing that these fundamental laws do not apply absolutely. This is accomplished by showing the construction of machines which can create or destroy energy and momentum.

CONTENTS

Chapter One

EQUALS

Perhaps the most fundamental of building blocks of mathematics is the concept of equals. The equals function in an equation states that the mathematical statement on one side of the equals sign is the same as the mathematical statement on the other side. In a non-mathematical concept, equals can also be used to signify that two things are the same. This chapter deals with the equality of lengths, or, perhaps, equality of length measurements is a more descriptive term. When two lengths are said to be equal, what is meant is that the measurement of each length is equal. Distinguishing between length and measurement of length may seem trivial, but the system used to do measurements is a characteristic of the mathematical system used to compare those measurements. Different measurement techniques will result in different measurements, even though the measurement technique does not physically alter the thing being measured.

For example, this chapter will first show that is it possible to have two different measurements for the length of the same line. The fundamental concept used in the discussion will be length contraction, one of the basic discoveries of the Special Theory of Relativity. This concept will be reviewed briefly to reacquaint the reader with it. The length contraction principle is closely linked with the behavior of time (the concept of

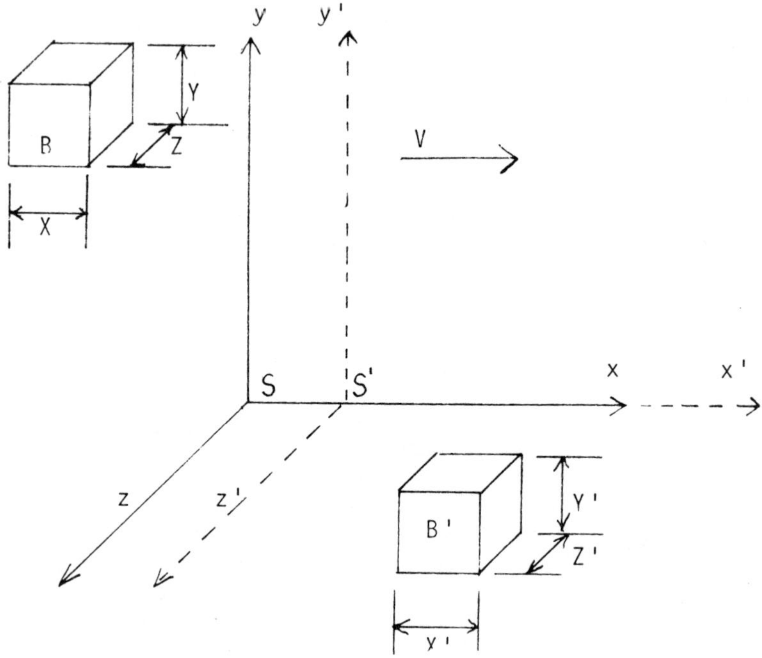

Figure 1

simultaneity, to be specific), and the reader should have some background in this relationship. For the first proof, consider the two coordinate systems of Figure 1. System S is stationary and system S' moves in a direction parallel to the x axis with a velocity V. Most good physics textbooks dealing with relativity present the same sort of picture with the derivation of the relationships of length in the two coordinate systems from the time dilation phenomenon. The results which have bearing here can be summarized using the two boxes shown; box B is at rest in system S and box B' is at rest in system S'. The edges of each box are parallel to the axes of its respective coordinate system according to the figure. The lengths of the sides of each box have been labeled by X, Y, Z, X', Y', and Z' according to the axis which that edge is parallel to in the figure. Both coordinate systems' axes are orthogonal with x, y, and z parallel to x', y', and z' respectively, and are inertial reference frames. When $V = O$,

$$X = X', \ Y = Y', \ Z = Z'. \tag{1}$$

The relationships derived by Einstein show the lengths of each edge in the S' system when observed from the S system.

$$X' = X\,(1 - V^2/C^2)^{\frac{1}{2}}, \ Y' = Y, \ Z' = Z \tag{2}$$

In brief, when an object moves past a stationary observer (in an inertial reference frame), the object appears shorter in the direction of motion. Note that there are no restrictions placed on X, Y, Z, X', Y', and Z'. They could easily represent infinitesimal lengths.

What about an object moving with an arbitrary motion relative to an inertial reference frame? An additional postulate of Special Relativity suggests that accelerated motion be dealt with as a series of instantaneous, uniform motions. Velocity influences measurements in an inertial reference frame, but acceleration does not. It is only necessary that system S of Figure 1 and equation *(2)* be inertial. System S' can be accelerating (with box B' stationary within it) and equation *(2)* still applies for an instant that velocity V exists.

These are the concepts needed for the proof. Consider the disc of Figure 2a with radius R and center O. To find a number approximately the circumference of the disc, a good method would be to draw a symmetrical figure of straight lines such as the hexagon in Figure 2b over the disc and add up the lengths of the sides of the figure as if they were placed end to end in a straight line.

$$Ci = Circumference \approx \sum_{i=1}^{6} Li \tag{3}$$

The notation $\approx$ means "approximately equal to." The answer obtained from such an exercise would vary depending on how the hexagon is drawn over the circle (the hexagon could conceivably be drawn completely inside the circle or completely outside the circle, for example). It can be shown that a figure of any number of sides can exactly represent the circumference of the circle if it is drawn properly. A figure with an infinite number of infinitesimally long sides (essentially, a locus of points) gives the correct length of the circumference using this procedure.

$$Ci = \lim_{n \to \infty} \sum_{i=1}^{n} Li \tag{4}$$

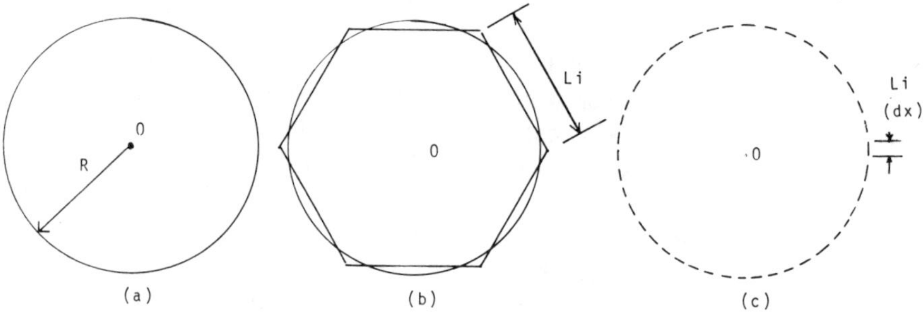

Figure 2

Equation *(4)* can be expressed as a definite integral. That integral may be interpreted as the line integral for arc length of the circumference. Replacing *Li* with *dx*:

$$C_{i} = \int_{c} dx = \int_{0}^{L} dx \tag{5}$$

Normally, a line integral would be taken to find *Ci* by reducing line segments to functions of a single variable. This case is already that simple. Because the line segments are assumed laid end to end in a straight line and then the sum taken, the equation *(5)* is the result sought. The function used here is just "1." That may seem like nonsense since the resulting equation $C_i = L$ doesn't say anything unless L is known, which means C_i is already known. The importance of *(5)* will be clearer later on. Note that L is the upper limit of the integral. Since the lower limit is zero, L is also the circumference of the circle.

$$C_i = L = 2\pi R \tag{6}$$

The circumference of the circle is also given by $2\pi R$ as is shown in *(6)*.

Now the above procedure will be repeated for a spinning disc. Assume Figure 2a is a disc spinning about O. The circumference may still be thought of as a line. If the angular velocity is $\vec{w}$, the tangential speed of any point on the line is $V = wR$. It is now desired to find the circumference of the disc while it is spinning. If, as before, an approximate circumference is desired, the disc could be approximated by a finite number of straight lines drawn over it, 1000 straight lines, for example. As before, when at rest, the circumference would be:

$$C_i = \sum_{1}^{1000} L_i \tag{7}$$

When the disc is spinning it would be approximately correct to assume that because $Li \ll R$, the small straight-line segments have a uniform velocity, V, everywhere over their lengths. The radius of the disc does not change due to relativistic effects because the radius is always perpendicular to the velocity of the circumference (which is a tangential velocity). Spinning the disc does not alter the number of straight-line segments from the number present when the disc is stationary.

An examination of an individual line segment is now made. The two coordinate systems of Figure 1 are placed instantaneously on it (see Figure 3). As before, system S is the observer's reference frame (inertial), and system S' is moving with velocity V for an instant with the straight-line segment stationary within it. The length of the straight-line segment when the disc is moving is:

$$L_i\,(1 - V^2/C^2)^{\frac{1}{2}} \tag{8}$$

The circumference of the disc while spinning is approximately:

$$C_i \approx \sum_{i=1}^{1000} L_i\,(1 - V^2/C^2)^{\frac{1}{2}} \tag{9}$$

Although it is not presented explicitly in Figure 3, every point along the length of the straight-line segment has a different velocity. This is due to the different lengths and orientations of radii which point to each point along that length. If the length *Li* is small compared to the radius, the differences in velocity along the length will be small. In fact, velocity becomes uniformly distributed over the line segment as its length approaches zero. This discussion is concerned only with the direction of the length of the line segment, since velocity in other directions doesn't influence the length.

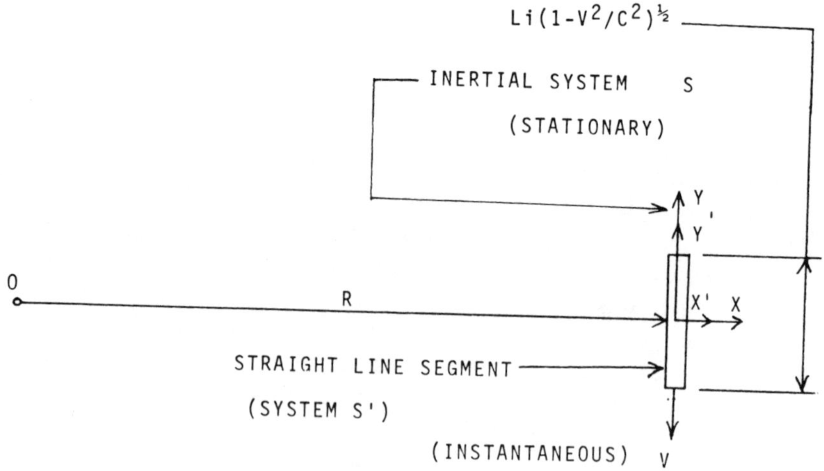

Figure 3

As was the case with the stationary disc, the circumference of the disc can be accurately found by assuming the figure drawn over it has sides which are infinitesimally long. That implies that the number of sides making up the figure must approach infinity. The circumference of the disc with an infinite number of sides is accurately found by *(4)* when at rest, so when the disc is spinning, the circumference is found by:

$$C_i = \lim_{n \to \infty} \sum_{i=1}^{n} L_i \left(1 - V^2/C^2\right)^{\frac{1}{2}} \tag{10}$$

Substituting dx for L_i, as before, *(10)* can be expressed as a definite integral.

$$C_i = \int_0^L \left(1 - V^2/C^2\right)^{\frac{1}{2}} dx \tag{11}$$

$$C_i = L\left(1 - V^2/C^2\right)^{\frac{1}{2}} \tag{12}$$

L has already been defined as the upper limit of the integral. When the disc is not spinning, $L = 2\pi R$. It will be necessary to show that L has not changed due to spinning the disc. In other words, L should not be a function of velocity as C_i appears to be. This can be shown by repeating the preceding process for the case of a straight line. Since the length of a straight line is already known in a stationary and moving reference frame, this will allow a determination of the nature of the upper limit of the integral of this procedure in general.

Suppose there is a line of length L_o when at rest. This length can be found by breaking the line up into equal-length segments, placing the segments end to end, and adding up the length of all of the segments. For a line stationary in an inertial reference frame, suppose, as before, the line is broken up into six equal-length sections and their lengths are added up.

$$L_o = \sum_{i=1}^{6} L_i \tag{13}$$

In this case, *(13)*, which is similar to *(3)*, gives an exactly correct answer. In fact, any number of equally long sections whose combined length is L_o will add up end to end to give the correct length of L_o, which is L_o. An infinite number of infinitesimally long segments will still be accurate.

$$L_o = \lim_{n \to \infty} \sum_{i=1}^{n} L_i \tag{14}$$

That is the definition of the definite integral.

$$L_o = \int_0^{L_s} dx \tag{15}$$

L_s is the upper limit of the integral in this case. Since the lower limit is zero, L_s is also the length of L_o.

$$L_o = L_s \tag{16}$$

Now it is assumed an observer is located in system S as before. The line being considered here is put on moving system S', with a velocity V in the direction of the length of the line. The length of the line when governed by relativistic effects is now L_r. If, when the line was stationary, it had been divided up into 1000 equally long segments L_i long, each segment would now be:

$$L_i\,(1 - V^2/C^2)^{\frac{1}{2}} \tag{17}$$

The length of the line while stationary would be:

$$L_o = \sum_{i=1}^{1000} L_i \tag{18}$$

The length of the line while moving would be:

$$L_r = \sum_{i=1}^{n} L_i \, (1 - V^2/C^2)^{\frac{1}{2}} \qquad (19)$$

Although *(19)* is an accurate statement of the length of the line, the length can still be expressed by the sum of infinitely many line segments, each infinitesimally long.

$$L_r = \lim_{n \to \infty} \sum_{i=1}^{n} L_i \, (1 - V^2/C^2)^{\frac{1}{2}} \qquad (20)$$

As before, that is the definition of the definite integral.

$$L_r = \int_{0}^{L_s} (1 - V^2/C^2)^{\frac{1}{2}} \, dx \qquad (21)$$

$$L_r = L_s \, (1 - V^2/C^2)^{\frac{1}{2}} \qquad (22)$$

Equation *(22)* is equivalent to equation *(12)*. L_s was equal to L_o when the line was stationary, and now L_s can be found when the line is moving because L_r is known when the line is moving with the help of *(2)*.

$$L_r = L_o \, (1 - V^2/C^2)^{\frac{1}{2}} \qquad (23)$$

When the line is moving, $L_s = L_o$. The upper limit of the integral is a constant in this case. Since the same procedure was followed for the previous case of the circumference of the circle, it must also be assumed that $L = 2\pi R$ holds when the disc is spinning, too.

Now it is possible to make a comparison of the computed circumference of the disc while stationary and while spinning.

$$C_i = L \qquad \text{stationary} \qquad (6)$$

$$C_i = L\,(1 - V^2/C^2)^{\frac{1}{2}} \qquad \text{spinning} \qquad (12)$$

The length contraction principle of the Special Theory of Relativity states that an object shortens in the direction of its velocity. Each section of the circumference of a circle has a velocity relative to an observer who is not spinning with the disc. The circumference must therefore appear shorter while in motion as a spinning disc. Yet, no change in radius has occurred. Therefore, if an observer looks at two discs side by side with one spinning, doesn't he see the same size circumference if both discs have the same radius?

Length is considered to be the distance between two points. The length of a straight line is the distance between the two points at the ends of the line. When the Theory of Relativity predicts that a straight line gets shorter in the direction of motion, it is not difficult to conceptualize what happens to the line. The endpoints get closer together. When a circle's circumference gets shorter, however, that is difficult to visualize when the radius stays constant.

The real problem here is language. If an observer sees a bar moving past him which has an initial length he knows to be L_o, he could put his ruler up quickly, measure the length while moving to be L (less than L_o), and conclude the bar is shorter than it was when it was at rest. But the bar has not changed length. The bar is still the same length as it always was. Instead, he can only conclude the bar measures less of a length while moving than while stationary. The hypothesis here is that the measurement of the bar's length and the actual length may not necessarily be the same (the actual length is the length the observer chooses to define in some reference frame). Measurement involves comparing an unknown quantity with a known quantity. A more mathematical statement would be "equating" the unknown to the known. The length of the bar L_o does not equal the measured length L. The expansion of the hypothesis is that the equals function is not directly applicable when comparing quantities in different reference frames.

The result is that a bar of length L_o, when stationary in an inertial reference frame and measured with a stationary ruler, is still L_o when the bar is moving, although the measured length L of the moving bar will be less than L_o when the ruler used is stationary in the inertial reference frame. In addition, the circumference of a stationary disc C_i is the same length when the disc is spinning. Both equations, *(6)* and *(12)*, still represent the measured circumference depending on how the measurement is made. If the observer places a clock on the spinning disc's circumference and multiplies the time as measured by that clock for one revolution of the disc times the speed of the circumference, he will see the result given by *(12)*. If he compares a like disc that is not spinning to the spinning one, he will note that discs that have the same dimensions while stationary also have the same dimensions when spinning. That result is given by *(6)*. The spinning disc example is important because the circumferences of spinning and stationary discs are obviously equal and not equal at the same time. The classical concept of "equals" is not satisfactory.

So a spinning disc has a circumference which is a line of a certain length when measured in one way, and another length when measured another way. Neither measurement method is incorrect, and both methods yield a number which accurately represents the circumference. If one thinks back to the bar of length L_o when at rest, and L when moving, neither length is an incorrect length; each length is just the number obtained from different measurement methods. Time also can be obtained with moving or stationary clocks, and neither number obtained by either method is "wrong," just different.

The statement that two different lengths can be correctly descriptive of the same line is, therefore, correct. The reason for that is that the notion of equality does not translate in a one-to-one relationship between moving reference frames. The principle of length contraction is not just a mathematical oddity. It complicates the entire mathematical system.

In order to further explain this phenomenon, reference will now be made to another situation. An observer is in an inertial reference frame watching a rocketship accelerating with a long bar in tow (Figure 4a). The length of the bar at rest will be L_o, and the length while moving will be L. At some instant, the bar will have velocity V, and according to *(2)* it will have an observed length:

$$L = L_o \, (1 - V^2/C^2)^{\frac{1}{2}} \tag{24}$$

Because the rocket is accelerating, the velocity V will be a function of time. Since L is a function of V, L will also be a function of time. The change in the bar's length as a function of time is:

$$V_i = dL/dt = \frac{L_o \; V/C^2}{(1 - V^2/C^2)^{\frac{1}{2}}} \, dV/dt, \tag{25}$$

with V and dV/dt always positive numbers. The term dL/dt will always be assumed to be positive as is indicated in *(25)* to simplify the mathematical manipulations in which it is to be used. The term dL/dt can be taken to be a velocity as is shown by *(25)*. It means that one end of the bar has a velocity toward the other end of the bar. If the end of the bar nearest the rocket has the same velocity as the rocket, V, then the end of the bar farthest from the rocket must have a velocity given by:

$$V' = V + V_i \qquad V' < C \tag{26}$$

This means that the end of the bar is moving faster than the rocketship. As a hypothesis, assume that the velocity of the end of the bar as measured by the observer is the true velocity of that part of the bar. From Special Relativity, it must be assumed that that velocity cannot exceed the speed of light. Combining equations *(25)* and *(26)* and solving for acceleration,

$$A_1 = \frac{dV}{dt} < \frac{C^2 (C/V - 1)(1 - V^2/C^2)^{\frac{1}{2}}}{L_o} \tag{27}$$

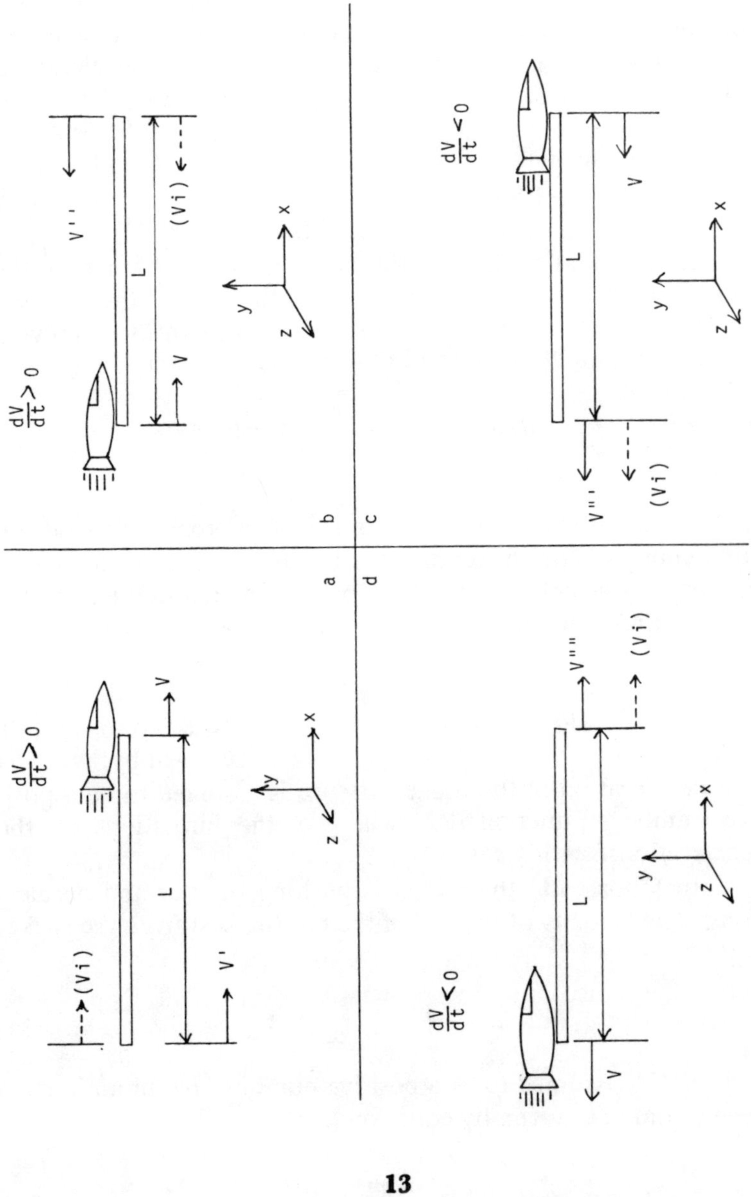

Figure 4

The maximum acceleration of the rocket is dependent upon the initial length of the bar and upon the velocity of the rocket.

Now refer to Figure 4b where the same rocket is pushing the same bar. If the rocket is again accelerating, the length of the bar will also be changing according to equation *(25)*. The velocity (assumed) of the end of the bar farthest from the rocket relative to the stationary observer is:

$$V'' = V - V_i \qquad - V'' < C \tag{28}$$

This means that the end of the bar is moving slower than the rocket in some cases. The velocity of the end of the bar will again be assumed not to exceed the speed of light, so, applying the same manipulations to *(28)* and *(25)*:

$$A_2 = \frac{dV}{dt} < \frac{C^2 (1 + C/V)(1 - V^2/C^2)^{\frac{1}{2}}}{L_o} \tag{29}$$

As before, the maximum acceleration of the rocket depends on the velocity of the rocket and the length of the bar. Shown in Figure 4c is a rocket with a long bar attached and the rocket is decelerating. The velocity of the end of the bar farthest from the rocket, $V''\,'$, is:

$$V''\,' = -V - V_i \qquad - V''\,' < C \tag{30}$$

The deceleration of the rocket, dV/dt, is assumed to be a positive number. Equation *(27)* will give the limitations on the deceleration for this case.

In Figure 4d, the rocket is pushing the bar and decelerating. The velocity of the end of the bar farthest from the rocket is V″ ″.

$$V''\,'' = -V + V_i \qquad V''\,'' < C \tag{31}$$

With dV/dt assumed to be a positive number, the limitations on deceleration are given by equation *(29)*.

Taking these four cases as real, it is apparent that rockets of the future will find accelerating much easier if they push long cargoes around. Much higher accelerations will be possible this way, while one end of the cargo is held to a velocity below the speed of light.

Actually, this awkward limit placed on the acceleration of objects is an example of the principle given previously. Equation *(24)* expresses the measured length of the bar. Hence *(25)* expresses a measured velocity of the end of the bar. One should wonder why there would be a different limit on acceleration depending on whether the rocket pushes the bar or pulls it.

To see what is really happening, look at the situation from the viewpoint of observers riding on the rocket. As was noted before, acceleration should be treated as a continuous series of instantaneous velocities. If one observer is on the rocket and another is on the far end of the bar, at any instant during the acceleration of the rocket, the situation of the observers is the same as if they were moving constantly at that instantaneous speed. If both observers are moving at a constant velocity, then they may be observed closer together by a third observer not in their reference frame [equation *(24)*], but they still see each other the same way as if they were not moving. The observers on the rocket and bar still measure themselves to be L_o distance apart no matter what continuous velocity they travel relative to a third observer. The logical extension of this idea for acceleration is that the contraction of the distance between the two moving observers as noted by a third observer not in their reference frame does not appear to the two observers on the rocket.

If the observers on the rocket measure the distance between them to be a constant during their acceleration, then they do not measure any velocity between them during the acceleration either. Since the observers on the rocket and far end of the bar do not measure any relative velocity between each other in their own reference frame, then they must both see a third observer in another reference frame as having the same velocity relative to each. If the observer on the rocket measures the velocity of the third observer to be V_1 and the observer on the far end of the bar measures the velocity of the third observer simultaneously to be V_2, then:

$$V_1 = V_2 \tag{32}$$

The conclusion that must be drawn is that every point on the rocket and bar must have the same velocity. This is a contradiction to *(25)*. Which philosophy is consistent with the present concepts of mathematics and physics?

Equations *(27)* and *(29)* place a limit on the acceleration of the rocket because one portion of the bar would supposedly go faster than the speed of light if that limit were exceeded. Suppose the rocket is considered to be stationary, however, and it is the third observer on another reference frame who is to be accelerating. Since the observer could be considered to be just a point, there would be no limitations on the acceleration that he could experience. If he exceeds the limits given by *(27)* or *(29)*, he would then observe one end of the bar to be going faster than the speed of light, even though he is moving and not the bar. It can be concluded, therefore, that it is not possible for an object to go faster than the speed of light, but it is possible for an object to be observed going faster than the speed of light. The measured velocity can be greater than the speed of light. Since it was crucial to the derivation of *(27)* and *(29)* that the measured velocity is the true velocity, the limitations presented by those equations do not exist.

The result of the discussion about the rockets is the same as the discussion about the spinning disc. The concept of equality does not translate directly from one reference frame to another. The velocity of the end of the bar of the accelerating rocketship can be measured to be faster than the speed of light. In other words, the velocity of the end of the bar can be equated to a known velocity which is greater than the speed of light, and cannot possibly be the actual velocity. The end of the bar could not go faster than the speed of light.

The equals function is the fundamental function of the mathematical system presently in use. The entire mathematical system relies on the equals function because it gives the system the ability to determine unknown quantities. The concept of equals involves comparison of quantities. One quantity is equal to another if they can be compared exactly.

The process of comparing one situation to another situation is measurement. Measurement can be used to determine unknown quantities because unknown quantities can be compared to known, or previously defined, quantities (standards, definitions, meters, etc.). One important aspect of these previously defined quantities is that they allow large numbers of situations to be grouped together in a precise notation (mathematical equations).

The two examples cited in this chapter show situations where the measurement of length and velocity are not compatible with the equals function. If the same length can have different measurements, the equals function is ambiguous. If a velocity which cannot exist can be observed (which implies it is measured in some manner), the equals function is ambiguous. The velocity cannot be "equal" to anything if it doesn't exist. The present mathematical system may describe these situations, but the equals function must be modified to include these situations.

Chapter Two

MASS AND ENERGY

The universe can generally be divided into two major constituents called matter and energy. Defining matter and energy is not easy, other than to say that they behave according to certain mathematical laws (the Law of Conservation of Momentum and Newton's Law, for example). One of the most significant statements of Relativity Theory is:

$$E = mc^2 \qquad (33)$$

This equation states that the energy, E, released by converting a mass, m, completely into energy is found by multiplying that mass by the speed of light, c, squared. The conversion of matter (mass) to energy is accomplished by nuclear fission. Equation *(33)* represents a relationship between energy and mass. It is a statement that energy and mass are interchangeable, one to the other and back again. That energy must have some relationship to mass can be further clarified by the following example.

Suppose there is a block with rest mass m_{b1} sliding across a frictionless surface with velocity $B = v/c$. The block is insulated so that heat inside it cannot escape. A mechanism inside the block converts some of the mass of the block into heat, which is held inside the block by the insulation. After the specified portion of the block's mass has been converted into heat, the matter remaining will have a mass m_{b2}, which is less than m_{b1}.

This simple process creates some difficulties with the most fundamental laws of physics, the Law of Conservation of Momentum and the Law of Conservation of Energy, when it is analyzed relativistically. Those laws state basically that energy/momentum cannot be created nor destroyed in an isolated system (no outside forces acting on the system). The laws are combined in relativistic calculations into a single law. The reasons for that combination will be reviewed.

According to relativity theory, all events are to follow the same fundamental laws when observed from any inertial reference frame. The Lorentz transformations describe how observations change from a reference frame which is moving relative to the event observed. When the Lorentz transformations are included into the fundamental laws describing an event, the results given by that law must be the same as the results obtained in the reference frame where the event is taking place; or the fundamental laws must be invariant under the Lorentz transformations. If x_1, x_2, and x_3 are three orthogonal coordinate axes in space, then the classical Law of Conservation of Momentum is:

$\triangle$ means "the change in" and n is the number of objects.

$$\triangle\left(\sum_{n} m \frac{dx_i}{dt}\right) = 0 \qquad i = 1, 2, 3 \tag{34}$$

Equation *(34)* is not invariant under the Lorentz transformations. The form of *(34)* which is invariant is:

$$\triangle\left(\sum_{n} m \frac{dx_i}{ds}\right) = 0 \qquad i = 1, 2, 3, 4 \tag{35}$$

$$ds^2 = c^2dt^2 - (dx_1^2 + dx_2^2 + dx_3^2) \tag{36}$$

In *(35)*, time is considered as part of the coordinate system, and is labeled dx_4. The term s is defined in relativity as the space-time interval, and is the new independent variable, replacing the previous independent variable of classical physics, time. If *(36)* is substituted into *(35)*, the result is:

$$\triangle \left[\sum_{n} \left(\frac{m \; c \dfrac{dx_i}{dt}}{(1 - B^2)^{\frac{1}{2}}} + \frac{mc^2}{(1 - B^2)^{\frac{1}{2}}} \right) \right] = 0 \qquad (37)$$

$$\vec{B} = \frac{\dfrac{\overrightarrow{dx_1}}{dt} + \dfrac{\overrightarrow{dx_2}}{dt} + \dfrac{\overrightarrow{dx_3}}{dt}}{c}$$

$$i = 1, 2, 3$$

One way of interpreting *(37)* is:

$$\triangle \left(\sum_{n} \left\{ \left[\frac{m}{(1 - B^2)^{\frac{1}{2}}} \right] c \frac{dx_i}{dt} + \left[\frac{m}{(1 - B^2)^{\frac{1}{2}}} \right] c^2 \right\} \right) = 0 \qquad (38)$$

$$i = 1, 2, 3$$

In *(38)*, terms have been arranged so that it appears as though relativistic momentum is of the same form as classical momentum, that is mass times velocity, $m'v$, where m' is the mass that is a function of velocity. This is where the notion of mass variance originates. There are certain advantages to thinking of mass as changing by this formula, but only in certain circumstances.

Equation *(37)* is the correct way of thinking of momentum and energy. It is directly derivable from *(35)*, and gives the Law of Conservation of Momentum and the Law of Conservation of Energy in the same equation. In simple terms, *(37)* says that the momentum and energy of an object are a function of its velocity. The energy and momentum, which *(37)* says are conserved, are represented in *(37)* as a function of the velocity of the object, B. The apparent variation of mass with velocity originates when the energy and momentum conservation laws are put in the form which is invariant under the Lorentz transformations. In general, the energy, E, and momentum, P, of an object are:

$$E = \frac{mc^2}{(1 - B^2)^{\frac{1}{2}}} \qquad P = \frac{mcB}{(1 - B^2)^{\frac{1}{2}}} \tag{39}$$

m = mass of object at rest
B = object velocity divided by c
c = speed of light

If the example of the block sliding on a frictionless surface is analyzed according to the classical momentum theory, the comparison of initial and final momentum is:

$$\begin{aligned} P_i &= m_{b1}v \quad \text{initial momentum} \\ P_f &= m_{b2}v \quad \text{final momentum} \\ P_i &\neq P_f \end{aligned} \tag{40}$$

The Law of Conservation of Momentum is contrary to this result. The more exact relativistic mathematics of *(37)* yields:

final (momentum + energy) − initial (momentum + energy) = 0

$$\frac{m_{b2}\,c^2B}{(1 - B^2)^{\frac{1}{2}}} + \frac{m_{b2}\,c^2}{(1 - B^2)^{\frac{1}{2}}} - \frac{m_{b1}\,c^2B}{(1 - B^2)^{\frac{1}{2}}} - \frac{m_{b1}\,c^2}{(1 - B^2)^{\frac{1}{2}}} = 0$$

$$m_{b2} = m_{b1} \tag{41}$$

Equation *(41)* states that *(37)* is applicable only if m_{b1} and m_{b2} are equal. Conceptually, a first instinct in this experiment is to suppose that since some of the block's matter is no longer present after it is converted into heat energy, the mass of the block, and hence its momentum, is reduced. The conflicts of

(40) and *(41)* indicate an error in this assumption. The statements of the Law of Conservation of Energy and the Law of Conservation of Momentum have been expanded in relativity theory to correct discrepancies such as this. This is accomplished by recognizing that the heat energy in the block is an integral part of the system. This energy must be included in the momentum calculations for the system. In classical physics, energy is characterized as having no mass, which has kept it from contributing to the momentum calculations of systems such as this. Claiming that energy has mass is difficult. Light is energy that travels at the speed of light. If light energy had mass, what is the interpretation of *(39)* with this velocity ($B = 1$)? A more philosophical argument is that, at the present time, there are still advantages to thinking of matter and energy as two different entities.

In order to overcome these difficulties, energy will not be characterized as having mass. It is only claimed that energy has "something," which shall be called "mass property," and which has the property of allowing momentum discrepancies of the type under discussion here to be accounted for using a specific calculation technique. That may seem like a vague definition; however, to define mass property is to try to characterize a property of energy as being similar to a property of mass, when there are no really good definitions of mass or energy. To explain the concepts of mass and energy requires a mathematical statement of fundamental laws which includes mass or energy in that statement. Newton's Law and the Law of Conservation of Energy are examples of the mathematical interpretation of mass and energy. Similarly, the notion of mass property will rely heavily on its relevance to mathematical statements of fundamental laws as its definition.

To characterize mass property mathematically requires that the Law of Conservation of Momentum calculations such as *(40)* and *(41)* be consistent with the concept. One requirement that must be fulfilled if this is to happen is that the center of mass property of the heat energy of the block be located in the same place in the block as the center of mass of the matter that was converted into the heat energy. The conversion of matter into energy should not alter the location of the center of mass of the block. To prove this, suppose an observer is sitting on the block as it slides along the frictionless surface. He knows some material in the block is being converted into energy. He knows the mass of the block plus the mass property of the energy can be added together to give a "total mass" of the block, $m(t)$, but the function $m(t)$ is unknown to him other than it is a finite number. He supposes the change of matter into energy will change the position of the center of total mass of the block. A reference frame is affixed to the block and distance r measures the position of the center of total mass relative to the reference frame. Since r is changing with time, it may be assumed that dr/dt, the rate of change of r with time, is available. Before the mass was converted into energy, the center of total mass velocity, dr/dt, was zero, so it must be concluded that the velocity also undergoes a change with time, d^2r/dt^2.

Newton's Law, one of the fundamental laws of classical physics, can be stated: the force applied to an object is equal to the rate of change of the object's momentum. Generally, Newton's Law is believed to apply to every physical situation. Newton's Law must be satisfied, then, for the object of this discussion. The net, or resultant, force acting on the system (the sum of all the individual forces acting on all of the parts of the system) results in a momentum change of the center of mass of the system. The general mathematical statement of Newton's Law is:

$$F = \frac{d\ (mv)}{dt} = m\frac{dv}{dt} + v\frac{dm}{dt} \qquad (42)$$

F = force
m = mass
v = velocity associated with m
t = time

The force acting on the block is zero. The velocity and acceleration of the center of total mass have been given. Plugging this information into *(42)*, Newton's Law at any instant of time during the conversion of mass to heat energy is:

$$0 = m\,(t) \left(\frac{d^2r}{dt^2}\right) + \frac{dr}{dt}\left(\frac{dm\,(t)}{dt}\right) \tag{43}$$

Even if $dm(t)/dt$ is zero, Newton's Law cannot be satisfied if the center of mass of the block is moving. The center of total mass property of the block must remain stationary during the conversion of mass to heat. Since the center of mass of the matter that is not converted into heat doesn't move during this process, the center of mass property of the heat energy must coincide with the center of mass of the matter that is converted.

Another requirement that can be placed on the mass property of the heat energy is that it must be equal to the mass of the matter that was converted. Suppose the mass property of the heat energy is m_h. The heat energy in the block can be found from *(33)*.

$$H = (m_{b1} - m_{b2})\ c^2 \tag{44}$$

The mathematical statement that the mass property of the heat is equal to the mass loss of the block is:

$$m_h = m_{b1} - m_{b2} \tag{45}$$

$$H = m_h\ c^2 \tag{46}$$

The above statements are not a derivation of the mass property of heat, but more like a subjective definition of what this property will have to contain if it is to correct the discrepancies in the previous momentum calculations. The point is that mass property is specifically related to another property of heat, that property being the heat energy. One of the basic revelations of the Theory of Relativity is that matter and energy are essentially the same thing [equation *(33)*]. Matter and energy are interchangeable. Matter can be converted into energy and energy

The ship has gone from a total momentum of zero to a momentum which is not zero. This means that the Law of Conservation of Momentum is not valid for this particular example. Since the Law of Conservation of Momentum is generally regarded to be a powerful law, it must be assumed that an error in the designer's philosophy exists. If the concept of mass property is introduced, the Law of Conservation of Momentum can be satisfied.

The mass property of the heat energy is to be considered as part of the spaceship. The heat energy is h, the mass property is m_h, and the definition of mass property is as previously discussed.

$$h = m_h c^2 \qquad \text{definition of mass property}$$

$$h = (m_{tb} - m'_{tb}) c^2 \qquad \text{amount of energy resulting from mass conversion}$$

$$m_h = m_{tb} - m'_{tb}$$

$$(m'_{tb} + m_h)v_{tb} = m_t v_t \text{ — ?} \qquad \text{moment of block and heat equals momentum of tube?}$$

$$[m'_{tb} + (m_{tb} - m'_{tb})]\, v_{tb} = m_t v_t$$

$$m_{tb} v_{tb} = m_t v_t \qquad (52)$$

The calculation procedure used previously for mass property is repeated for the case of the ship in *(52)*. The momentum of the ship will be zero in Figure 5c because the total momentum of the block (material and energy momentum) will be equal to the momentum of the tube. If the conversion of mass to energy did not take place, the tube would experience a force due to this mass when the relative block-tube velocity is brought to zero. With that mass converted into energy, the tube must experience the same force as the relative block-tube momentum is similarly brought to zero, or the Law of Conservation of Momentum will be contradicted. The total mass of the block (with energy included) is a constant throughout the process of Figure 5. Equations *(50)* and *(51)* are in error. The ship goes nowhere.

Mass property is a concept that can be extended to other forms of energy. The energy of electromagnetic radiation can also be shown to follow the principles of the discussion on heat energy. Light is a portion of the total spectrum of electromagnetic radiation. This total spectrum is the subject of this discussion but, for brevity, light will be the name used to represent the total spectrum. In the previous examples, heat energy was shown to react to forces as matter does. One way of producing light is to raise the temperature of certain materials. When these materials get hot enough, the heat energy is changed into light energy. There is a process available, therefore, whereby matter may be changed into light energy (with heat energy as an intermediate step).

Consider a hypothetical universe with two planets in it. The planets are both stationary in an inertial reference frame. These planets form a system, and, because they are the only objects in this universe there can be no forces external to this system acting on the system. Planet *1* will be assumed to exist emitting no light up until the time designated for this experiment to begin. It then emits light continuously for a "short" period of

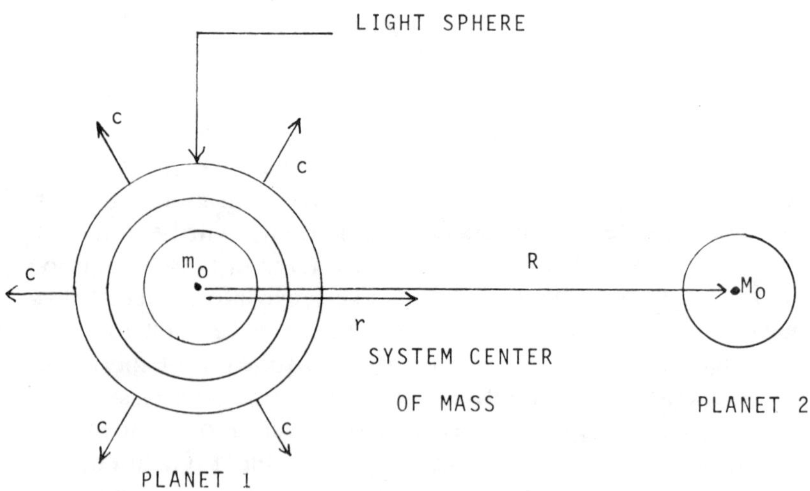

Figure 6

time, and then is completely dark again. The light emitted during this period instantaneously begins (no transients) and continues with all of its properties being uniform in all directions before it instantaneously stops. The light forms a spherical "shell" about planet *1*, which is at the geometric center of the light sphere at all times. The radius of the light sphere increases at the speed of light. The thickness of the light sphere wall is uniform.

Another planet (called planet 2) is positioned as shown in Figure 6. The two planets are uniform in composition and are spherical so that the center of mass lies at the geometric center of the planet. If a radius vector $\vec{R}$ designates the position of the center of mass of planet *2* with respect to that of planet *1*, then the initial center of mass of the planet *1*–planet *2* system must lie on the line of the vector $\vec{R}$ between the two planets, as is shown in Figure 6. If the mass of planet *1* is m_o and the mass of planet *2* is M_o, then the vector $\vec{r'}$ shows the position of the center of mass of the system. Vectors will be designated as a magnitude r' times a unit vector $\hat{r}'$ which gives the proper direction of the vector and has a magnitude of one.

$$\vec{r'} = r'\hat{r}' \qquad \vec{R} = R\hat{R} \qquad \hat{r}' = \hat{R}$$

$$r' = \frac{\mathrm{R}}{(1 + m_o/M_o)} \tag{53}$$

These equations assume that planet *1* is at the origin of the coordinate system. Gravitational forces between the two planets will be assumed negligible so that r' and R are constant with time.

The light emitted by planet *1* is the result of the conversion of some of the planet's material (mass) into light energy. The conversion of planet *1* mass to light energy will result in the total planet *1* mass being dependent on time. The total mass of the planet while the light is being emitted is $m(t)$ and the final mass is m_a.

$$m_a < m_o \tag{54}$$

Because the mass of planet *1* has changed, the center of mass of the planet *1*-planet *2* system is changing in magnitude and location. The new center of mass will be designated by r_a and the transient center of mass will be designated by r.

$$r = \frac{R}{(1 + m(t)/M_o)}$$

$$r_a = \frac{R}{(1 + m_a/M_o)} \qquad \hat{r} = \hat{r}' = \hat{r}_a \tag{55}$$

If the period of time planet *1* is emitting light is considered exclusively, the rate of change of the position of the center of mass of the system is:

$$\frac{dr}{dt} = -\frac{R/M_o}{(1 + m(t)/M_o)^2}\frac{dm(t)}{dt} \tag{56}$$

For simplicity, the rate of change of total mass of planet *1* into light will be assumed to be a constant. The acceleration of the center of mass of the system is:

$$\frac{dm(t)}{dt} = \text{constant} \qquad \frac{d^2m(t)}{dt^2} = 0$$

$$\frac{d^2r}{dt^2} = \frac{2R/M_o^2}{(1 + m(t)/M_o)^3}\left(\frac{dm(t)}{dt}\right)^2 \tag{57}$$

The force acting on the planet *1*–planet *2* system is zero, from the previous description of the system. The velocity and acceleration of the center of mass of the system are given by *(56)* and *(57)* respectively. The mass of the system at any instant of time during the light emission period is $m(t) + M_o$. Plugging this information into Newton's Law, *(42)*, which must be true for this system:

$$0 = [m(t) + M_o]\left[\frac{2R/M_o^2}{(1 + m(t)/M_o)^3}\right]\left(\frac{dm(t)}{dt}\right)^2 - \frac{R/M_o}{(1 + m(t)/M_o)^2}\left(\frac{dm(t)}{dt}\right)^2$$

$$0 = \frac{R/M_o}{(1 + m(t)/M_o)^2}\left(\frac{dm(t)}{dt}\right)^2 \qquad (58)$$

Equation *(58)* is not a valid equation. Newton's Law is not satisfied. Even though the planets are not moving, the center of mass of the system is changing, resulting in *(58)*. Accounting for the discrepancies presented by this experiment is accomplished by recognizing that the light emitted by planet *1* is an integral part of the system. Light will be assigned mass property, defined similarly to the mass property of heat energy, and the Newton's Law calculation will be repeated.

The magnitude of the mass property of the light, m_l, is to be equal to the mass of planet *1* which has been converted into the light energy. In this way, the mass of the planet *1*–light subsystem is always a constant. The light energy is designated by *E*.

$$E(t) = [m_o - m(t)]c^2$$

$$m_1(t) = m_o - m(t)$$

$$E(t) = m_1(t)c^2 \qquad (59)$$

In *(59)*, the quantities involved are represented as a function of time since this situation is transient. The above discussion is not a derivation of the mass property of light, but more like a subjective definition of what this property will have to contain if it is to correct the discrepancies in the previous Newton's Law calculation. By the above reasoning, mass property is specifically related to another property of light, that property being the light's energy.

The center of mass of the system will be determined, considering the light as part of the system. The mass of planet *2* is a constant in this experiment. If the center of mass of the system is to be a constant in magnitude and location, the center of mass of the planet *1*–light subsystem must be a constant. If, at any point in time, the total energy emitted by planet *1* in the form of light is $E(t)$, then the mass of planet *1* will be $m_o - E(t)/c^2$. At any point in time, the mass property of the emitted light will be $m_1(t) = E(t)/c^2$.

If the center of mass of the planet system, including the light, is to remain stationary during and after the light emission process, the center of mass property of the light must coincide with the center of mass of planet *1*. In the example of Figure 6, this is the case if mass property is defined properly. To define mass property intuitively, it must be related to some other property of light. It was stated earlier that all the properties of the light are uniform in all directions, including the geometric form of the volume of space instantaneously occupied by the light (the spherical shell describing where the light is at any point in time). If mass property is related to some other property of light, then the mass property of the light forming the spherical shell around planet *1* with planet *1* at the geometric center of the sphere is distributed spherically around planet *1* with the center of mass property being at the geometric center of the sphere. The center of mass property of the light is in exactly the same place as the center of mass of planet *1*, according to the previous definitions of the system. As an analogy, if the spherical shell of light were actually matter (i.e., a spherical balloon inflating with planet *1* at the geometric center of the balloon), the center of mass of the spherical shell of matter would coincide with the center of mass of planet *1* because of the positioning of one relative to the other. The same logic is applied for the position of the center of mass property of the spherical shell of light relative to the center of mass of planet *1*. The mass of the planet *1*–light subsystem, m_c, at any point in time will be the addition of the two masses, and the location of the resultant total mass will be at the location of the planet *1* mass before the light emission.

$$m_c = [m_o - E(t)/c^2] + m_1 = m_o \qquad (60)$$

The total mass of the planet *1* light subsystem, m_c, is a constant. The center of mass of the planet *1* planet *2* light system will also be a constant. The variation of mass of the system when considered as a whole is zero with time. The variation of the position of the center of mass has to be zero in this case if the mass of the system is a constant. Newton's Law will be zero equals zero, and will be satisfied. Newton's Law will be intuitively satisfied. There is no exterior force on the system, so the position of the center of mass will not change. Note that the mass property of the light beam is not presumed to change with velocity. The mass of objects is sometimes considered a function of velocity, due to the grouping of terms in the momentum equation of *(38)*.

$$m = \frac{m_o}{(1 - B^2)^{\frac{1}{2}}} \qquad (61)$$

m—apparent mass of moving object
m_o—rest mass of object
B—velocity of object

As before, this reasoning is useful only in limited circumstances. The mass property of light is not dependent on velocity.

As with light and heat energy, kinetic energy can also be shown to have mass property. This can be shown using the machine of Figure 7. This machine is a block of mass m_1 which is traveling within a framework with a mass m_2. The framework is sliding across a surface with velocity B_y. The velocity of the block relative to the framework is B_x. All surfaces are frictionless surfaces, and this experiment takes place in a universe where there are no other objects to exert gravitational effects on

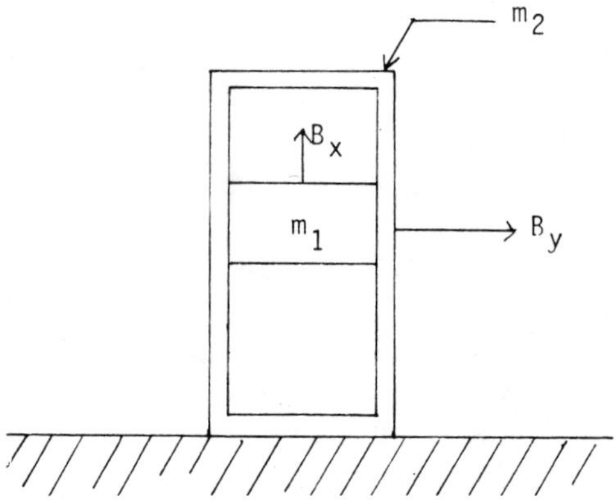

Figure 7

the machine. The gravitational effects within the machine are negligible. The momentum of the machine in the direction of B_y will be determined. If B_x is zero, the momentum of the machine will be:

$$\overline{P} = \frac{(m_1 + m_2)c\overline{B}_y}{(1 - B_y^2)^{\frac{1}{2}}} \tag{62}$$

If B_x is not zero, the momentum of the m_1 block is defined in terms of the block's total velocity, B_t.

$$\overrightarrow{P'} = \frac{m_1 c\overrightarrow{B_t}}{(1 - B_t^2)^{\frac{1}{2}}} \tag{63}$$

As before, B_t is the magnitude of vector $\overrightarrow{B_t}$. It is necessary to find B_t in terms of B_x and B_y. From relativity theory, the transformation of velocities from one reference frame to another can be found. In Figure 8, an object has velocity u' in the primed inertial reference frame which moves with velocity v relative to the unprimed reference frame. The direction of v is in the direction of the y' and y axes, which are parallel. The x' and x axes are parallel, and the z', and z axes are parallel. The velocity u'

has components u'_x, u'_y, and u'_z along the x', y' and z' axes respectively. The corresponding velocities seen by an observer on the unprimed frame are:

$$u_y = \frac{u'_y + v}{1 + \frac{vu'_y}{c^2}} \qquad u_x = \frac{u'_x(1 - v^2/c^2)^{\frac{1}{2}}}{1 + \frac{vu'_y}{c^2}} \qquad u_z = \frac{u'_z(1 - v^2/c^2)^{\frac{1}{2}}}{1 + \frac{vu'_y}{c^2}} \tag{64}$$

In the case of Figure 7, the moving primed reference frame will be the framework, m_2, and the object with velocity u′ will be the m_1 block. The unprimed stationary reference frame will be

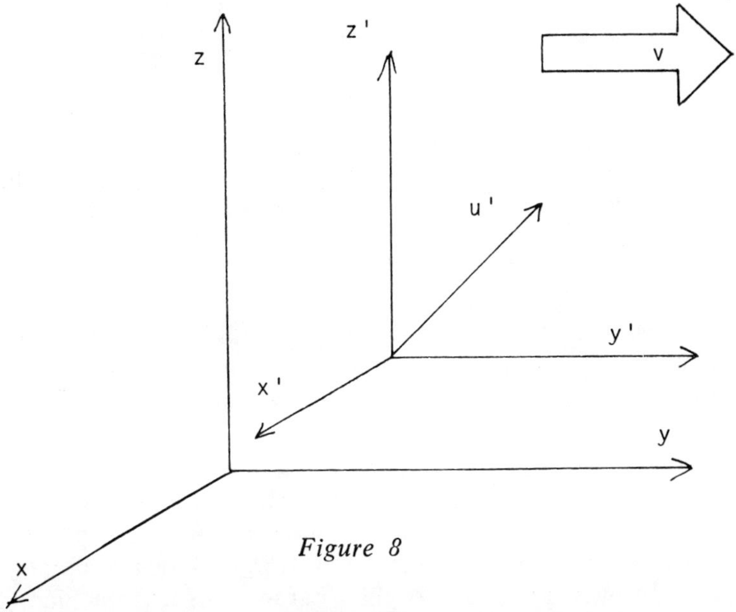

Figure 8

the frictionless surface m_2 is sliding on. The vector B_t, according to *(64)* is:

$$v = v_y = cB_y \qquad u'_x = v_x = cB_x \qquad u'_y = 0 \qquad u'_z = 0$$

$$\overrightarrow{B_t} = \frac{\pm v_x(1 - v_y^2/c^2)^{\frac{1}{2}}}{c}\,\hat{i} + \frac{v_y}{c}\,\hat{j} \tag{65}$$

$\hat{i}, \hat{j}$ = unit vectors in the x, y directions respectively

The $\pm$ sign indicates that the direction of the motion of the m_1 block can be reversed. From vector calculus, the magnitude of B_t is:

$$B_t^2 = \frac{(1 - v_y^2/c^2)v_x^2}{c^2} + \frac{v_y^2}{c^2} = B_x^2 - B_x^2 B_y^2 + B_y^2 \qquad (66)$$

Plugging *(65)* and *(66)* into *(63)*, the momentum in the B_y direction for the whole machine is:

$$P'' = \frac{m_1 c B_y}{(1 - B_x^2 - B_y^2 + B_x^2 B_y^2)^{\frac{1}{2}}} + \frac{m_2 c B_y}{(1 - B_y^2)^{\frac{1}{2}}}$$

$$P'' = \frac{\left[\dfrac{m_1}{(1 - B_x^2)^{\frac{1}{2}}} + m_2\right] c B_y}{(1 - B_y^2)^{\frac{1}{2}}} \qquad (67)$$

The total mass of this machine will be needed to find the machine momentum. The mass property of the kinetic energy will contribute to this momentum in the same way that heat energy mass property contributed to the momentums of previous examples. If the m_1 block had heat energy instead of kinetic energy and the mass property of the heat energy is m_h, the total mass of the machine in the moving reference frame affixed to m_2 would be $m_1 + m_2 + m_h$. Since the m_1 block has kinetic energy of mass property m_K, the total mass of the machine in the moving reference frame is $m_1 + m_2 + m_K$. The fact that m_1 is moving relative to m_2 is not given mathematical significance in the expression for the mass of m_1. The mass of m_1 relative to the reference frame of m_2 is not $m_1/(1 - B_x^2)^{1/2}$. The Theory of Relativity does not predict that the mass of the block's material increases as is often interpreted according to *(38)*. Equation *(37)* is the correct statement of the Law of Conservation of Momentum mathematics. The mass of the material of the m_1

block is still m_1, whether the block is moving or not, but the "total mass" of the block increases when the mass property of the kinetic energy is added.

Using the above expression for total mass of the machine, the assumption can be made that kinetic energy does not have an effect on the total machine momentum. The momentum of the machine in this case would be:

$$P = \frac{(m_1 + m_2)cB_y}{(1 - B_y^2)^{\frac{1}{2}}} \tag{68}$$

Equation *(68)* is equivalent to *(62)*, which is the relativistic momentum expression for $B_x = O$. If $B_x = O$, there is no kinetic energy effect in the momentum. But the assumption is that *(68)* is true for any value of B_x, since kinetic energy should have no effect on momentum no matter what its value. The momentum for any value of B_x, has already been determined in *(67)*, which does not agree with *(68)*. Apparently, the assumption must be made that kinetic energy does contribute to the machine momentum.

$$P = \frac{(m_1 + m_2 + m_K)\, cB_y}{(1 - B_y^2)^{\frac{1}{2}}} \tag{69}$$

Kinetic energy mass property will be defined in a manner similar to the way heat and light energy mass property is defined. Note that m_1, m_2, and m_K are all rest masses in the reference frame attached to m_2.

$$E_K = \frac{m_1c^2}{(1 - B_x^2)^{\frac{1}{2}}} - m_1c^2$$ m_1 kinetic energy from reference frame on m_2

$$m_K = \frac{m_1}{(1 - B_x^2)^{\frac{1}{2}}} - m_1$$ kinetic energy mass property relative to m_2 reference frame

$$P = \frac{\left[\dfrac{m_1}{(1 - B_x^2)^{\frac{1}{2}}} + m_2\right]cB_y}{(1 - B_y^2)^{\frac{1}{2}}} \tag{70}$$

If the kinetic energy mass property is added into *(69)* and defined as it was for other energy types, *(70)* is the result. By comparing *(70)* with *(67)*, it can be seen that the addition of mass property in this manner to the momentum equation of the machine agrees with the momentum expression obtained by using the velocity transformations of *(65)*.

The idea that energy reacts in some ways as mass does is a difficult idea to conceptualize, especially for kinetic energy. A convincing argument that can be presented to support this theory is the proof that potential energy also has mass property. The following discussion shows that the potential energy of a compressed spring makes it react differently inertially from an unstressed spring. It is not difficult to substitute any form of potential energy for that of the spring, once the theory of this proof is understood, however.

A machine will be constructed in the following discussion, and its parts will be shown separately at first. The final machine will gradually be built up so that its operation is easily seen. The first part to be considered is a block sliding across a frictionless surface. The mass of the block at rest is m_1 and the momentum of the block at velocity B_x is:

$$P_1 = \frac{m_1\, cB_x}{(1 - B_x^2)^{\frac{1}{2}}} \qquad (71)$$

Figure 9a shows a pair of identical blocks in a framework of frictionless surfaces. Each block is guided in its path by a straight track (not shown). The velocity vectors of the blocks are equal at all times, except that the velocities are always in opposite directions. The tracks are the same length, so each block travels the same distance. With the specification on velocity given, this means each block will be compressing the center springs or the springs at the end of the framework at the same time. The blocks will appear to be mirror images of each other. The four springs shown are identical in all respects and are positioned at the end of each block's travel as shown. This arrangement will be chosen because the momentum of each block in the

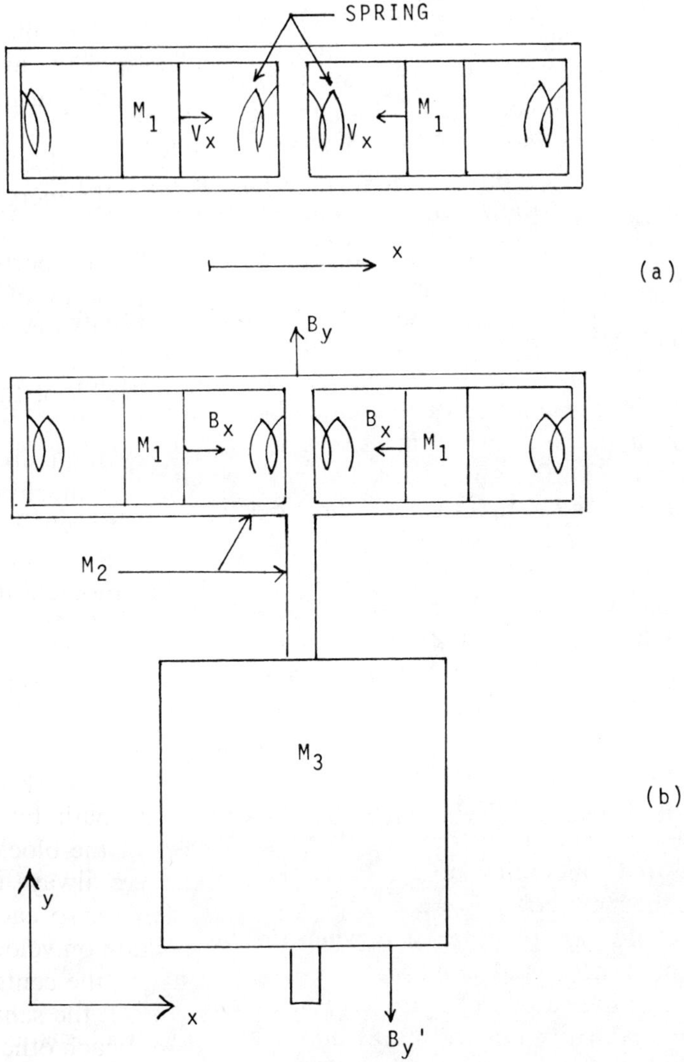

Figure 9

x direction will be the opposite of the momentum of the other block. The forces generated by any spring when a block runs into it will be equal but opposite to the force generated by one of the other springs at the same time. If this mechanism were stationary or moving with a constant velocity in free space, the motion of the two blocks labeled m_1 would not cause the framework to have any motion. The blocks would continue to travel back and forth in their tracks, rebounding off the springs indefinitely, but the framework would not show any movement or transmit any force to anything that would grab onto the framework.

In Figure 9b, the framework is attached to the end of a pole which has one end located at the center of mass of the blocks-framework system. The mass of the framework and pole is m_2. The pole extends through another body with mass m_3, and passes through the center of mass of that body. The pole slides on frictionless surfaces inside m_3 that may be considered to be closely toleranced so that the pole can move with a velocity only in the y direction indicated on Figure 9b.

The m_1-m_2-m_3 system is located in free space away from the influence of other objects. It will be assumed that there is some mechanism in the framework that can grab both blocks simultaneously at either end of their tracks at the instant when their velocity relative to the framework is zero. This also represents the extreme positions of each block in the tracks, and the point of maximum compression of the springs. This mechanism can be operated when desired. There is also a mechanism in the mass m_3 that can move the pole running through it with any force desired, including no force, or the state of free motion (only in the y direction, however).

The situation of interest is when the m_1-m_2 system has velocity $v_y = cB_y$, and mass m_3 has velocity $v'_y = cB'_y$, with directions as indicated in Figure 9b. Each block m_1 has velocity v_x (or $-v_x$) relative to the framework m_2 with their directions of motion still being opposite. The blocks' positions are somewhere in the middle of the tracks where they are not touching any springs. The momentums of m_2 and m_3 in this situation are:

$$P_2 = \frac{m_2 c B_y}{(1 - B_y^2)^{\frac{1}{2}}}$$

$$P_3 = \frac{m_3 c B'_y}{(1 - B'^2_y)^{\frac{1}{2}}} \qquad (72)$$

The momentum of each m_1 block is defined in terms of its total velocity, B_t, with respect to the $x-y$ reference frame. This momentum is given in *(63)*, as this machine is similar to the one illustrating the mass property of kinetic energy in this respect. The discussion of B_t in terms of B_x and B_y also applies, with *(66)* being the result in this case too. From the discussion of Figures 9a and 9b, the momentum of the system in the x direction will be zero. It will also be assumed that the velocities v_y and v'_y are such that the momentum Q_1 in the y direction is zero. The momentum in the y direction of the two blocks and framework is equal to the momentum of the mass m_3 in the negative y direction.

$$2P_{1y} + P_{2y} - P_{3y} = Q_1 = O$$

$$2\left[\frac{m_1 c B_y}{(1 - B_x^2 - B_y^2 + B_x^2 B_y^2)^{\frac{1}{2}}}\right] + \frac{m_2 c B_y}{(1 - B_y^2)^{\frac{1}{2}}} - \frac{m_3 c B'_y}{(1 - B'^2_y)^{\frac{1}{2}}} = Q_1 = 0 \qquad (73)$$

The blocks with mass m_1 will eventually come to the ends of the tracks and will run into the springs. They will be slowed down simultaneously until their velocity relative to the framework of mass m_2 is zero. The total momentum of the system in the x direction remains zero throughout this process. The two blocks are then grabbed when their velocity is zero by the mechanism mentioned before and held stationary relative to the framework. The velocity of each block is now v_y in the y direction and the momentum of each block is:

$$P_1^* = \frac{m_1 c B_y}{(1 - B_y^2)^{\frac{1}{2}}} \qquad (74)$$

The total momentum of the system in this situation is:

$$2P^*_{1y} + P_{2y} - P_{3y} = Q_2$$

$$2\left[\frac{m_1 c B_y}{(1 - B_y^2)^{\frac{1}{2}}}\right] + \frac{m_2 c B_y}{(1 - B_y^2)^{\frac{1}{2}}} - \frac{m_3 c B'_y}{(1 - B'^2_y)^{\frac{1}{2}}} = Q_2 \qquad (75)$$

The total momentum of the system Q_2 is not equal to the total momentum of the system Q_1. There have been no external forces on the mechanism shown, only the internal forces exerted by the springs and the mechanism grabbing the m_1 blocks. Since the initial system momentum was zero and no external forces have been applied, the final system momentum must also be zero.

The mass property of potential energy will now be considered in this situation. First, the process of a block sliding into a spring will be examined. In Figure 10a, a block is sliding across a frictionless surface toward a spring. According to relativity theory, it has momentum given by *(71)*.

$$P_1 = \frac{m_1 c B_x}{(1 - B_x^2)^{\frac{1}{2}}} \qquad (71)$$

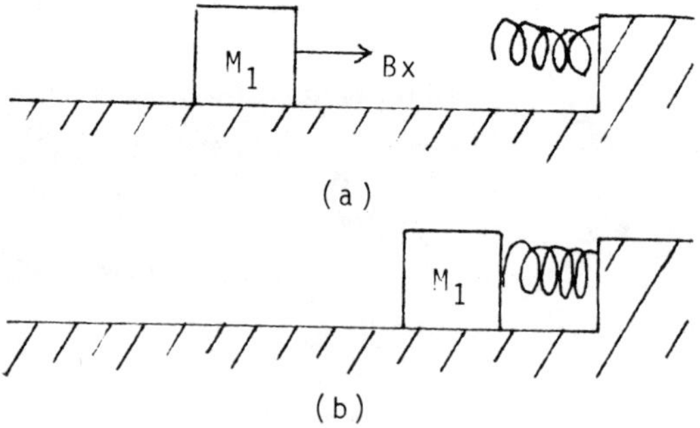

Figure 10

From the previous discussion of this chapter, the total mass of the block is the rest mass, m_1, plus the mass property of the kinetic energy of the block, given by dividing the kinetic energy by c^2.

$$m_p = \frac{\dfrac{m_1c^2}{(1 - B_x^2)^{\frac{1}{2}}} - m_1c^2}{c^2} = \frac{\text{kinetic energy of block}}{c^2} \qquad (76)$$

In the previous example on heat energy, the mass property of the heat energy was treated exactly as mass when the momentum of an object was calculated. In *(39)*, the mass in the momentum equation is the mass of an object in the frame at which it is at rest. The momentum is a function of the velocity and mass as shown. The momentum of the heat mass property is used identically [see *(47)* and *(48)*]. The rest mass property of potential energy will be used in the same way.

In Figure 10a, the mass property of the kinetic energy plus the rest mass of the block give a total mass for the block of $m_1/(1-B^2_x)^{½}$. In Figure 10b, the block has run into the spring and is at the instant that its velocity is zero. The mass of the block is m_1 in the stationary situation. The mass property of the potential energy of the spring is the potential energy stored in the spring divided by c^2. The potential energy stored in the spring, however, is equal to the kinetic energy of the block before it touched the spring, from the Law of Conservation of Energy. The mass property of the coiled spring will also be given by *(76)*. The mass of the block and mass property of the coiled spring can be expected to react to Newton's Law in the same way. In the stationary reference frame illustrated in Figure 10, the total mass of the block and spring is the same at all times throughout the process where the block slides into the spring.

Figure 11a shows the mechanism of Figure 9b with the masses m_1 held at the extreme positions in their tracks by the grabbing mechanism. The two springs shown are fully compressed. As was just shown, this is equivalent to the situation of Figure 11b where there is no coiled spring, but mass m_{1s} instead.

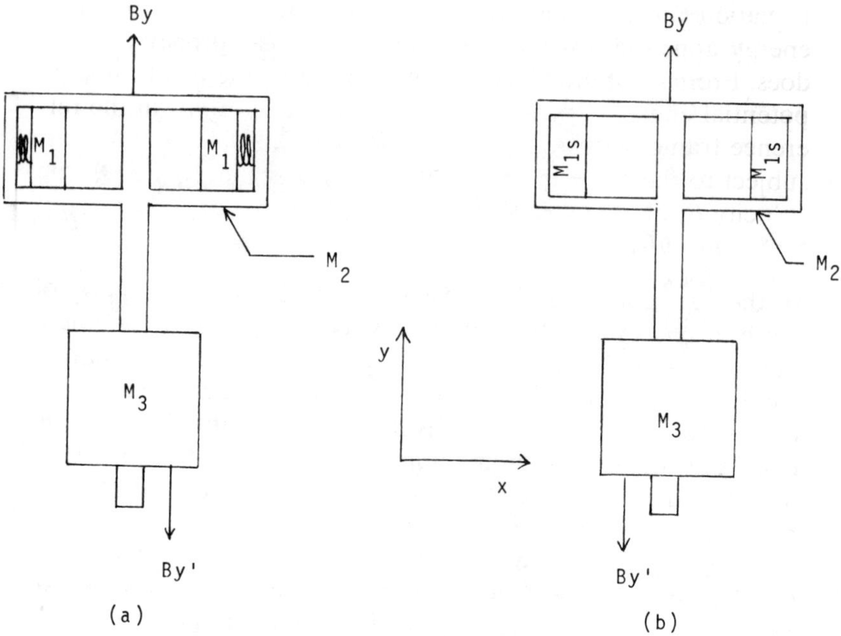

Figure 11

$$m_{1s} = m_1 + m_p = \frac{m_1}{(1 - B_x^2)^{\frac{1}{2}}} \qquad (77)$$

The momentum of the mass m_{1s} in the y direction is:

$$P_{1s} = \frac{m_{1s} c B_y}{(1 - B_y^2)^{\frac{1}{2}}} = \frac{\dfrac{m_1}{(1 - B_x^2)^{\frac{1}{2}}} c B_y}{(1 - B_y^2)^{\frac{1}{2}}}$$

$$= \frac{m_1 c B_y}{(1 - B_x^2)^{\frac{1}{2}} (1 - B_y^2)^{\frac{1}{2}}} = \frac{m_1 c B_y}{(1 - B_x^2 - B_y^2 + B_x^2 B_y^2)^{\frac{1}{2}}} \qquad (78)$$

Equation *(78)* uses the fact that the mass property of potential energy appears to vary in the same way with velocity as mass does. From the discussion on kinetic energy, this means that the potential energy contributes to the total mass of m_{1s} in the reference frame where the spring is at rest, and that total mass is subject to the gains in kinetic energy mass property due to the velocity B_y of the total mass. The total mass of the m_1 block and spring in the *x-y* reference frame is a combination of the mass of the block's material, the potential energy mass property, and the kinetic energy mass property.

Since m_{1s} represents the mass of the stationary block m_1 and the mass property of the coiled spring, *(78)* is also the momentum of those two components. P_{1s} is equal to P_{1y}, however. If the mass property of the potential energy of the coiled spring is considered, the total momentum of the system of Figure 11 is:

$$2P_{1s} + P_{2y} - P_{3y} = Q_3$$

$$P_{1s} = P_{1y}$$

$$2P_{1s} + P_{2y} - P_{3y} = Q_1$$

$$Q_1 = Q_3 \qquad (79)$$

The final momentum in the *y* direction is equal to the initial momentum, and the Law of Conservation of Momentum is satisfied.

As another example, an examination of the system of Figure 12 will be made. In Figure 12a, two masses, m_1 and m_2, have velocities of B_1 and B_2 such that the blocks approach each other. The mass labeled m_1 has a spring attached to it. The mass labeled m_2 has a latching mechanism such that, when the two masses come together against the spring and compress it to the point where their velocity relative to each other is zero, the latching mechanism will operate and hold the two blocks fixed in this position. As before, the spring is at its maximum compression of the example in this situation. The blocks will be closer together than at any point in time and the potential energy of the spring will be at its maximum for this experiment.

After the latching mechanism has operated, the situation will be as shown in Figure 12b. The two blocks will now be moving as one unit with velocity B_3. The momentums of the situations of Figure 12, without the concept of spring potential energy mass property, will be:

$$P_1 = \frac{m_1 c B_1}{(1 - B_1^2)^{\frac{1}{2}}} \qquad P_2 = \frac{m_2 c\,(-B_2)}{(1 - B_2^2)^{\frac{1}{2}}}$$

blocks moving independently in Figure 12a

$$P_3 = \frac{(m_1 + m_2)\, c B_3}{(1 - B_3^2)^{\frac{1}{2}}} \tag{80}$$

blocks moving as one unit in Figure 12b

If a primed reference frame moving with velocity B_3 is considered, the coupled blocks of Figure 12b will have a momentum of zero with respect to this reference frame. The total momentums of the two individual blocks must also add up to zero for the primed reference frame, as is shown in Figure 12c. The velocities of the two blocks relative to the primed reference frame are B'_1 and B'_2. The momentums of the two blocks relative to the primed reference frame are:

$$P'_1 = \frac{m_1 c\, B'_1}{(1 - B'^2_1)^{\frac{1}{2}}} \qquad P'_2 = \frac{m_2 c\,(-B'_2)}{(1 - B'^2_2)^{\frac{1}{2}}}$$

$$P'_1 = -P'_2 \tag{81}$$

According to *(64)*:

$$B_1 = \frac{B'_1 + B_3}{1 + B'_1 B_3} \qquad -B_2 = \frac{-B'_2 + B_3}{1 - B'_2 B_3} \tag{82}$$

If *(82)* is inserted into *(80)*, the result is:

$$P_1 = \frac{m_1 c\,(B'_1 + B_3)}{(1 - B'^2_1)^{\frac{1}{2}}\,(1 - B^2_3)^{\frac{1}{2}}} \qquad P_2 = \frac{m_2 c\,(-B'_2 + B_3)}{(1 - B'^2_2)^{\frac{1}{2}}\,(1 - B^2_3)^{\frac{1}{2}}} \qquad (83)$$

Equation *(83)* is a statement of the momentum of the blocks in the unprimed reference frame in terms of the velocities of the blocks in the primed reference frame and B_3. If P_1 and P_2 of *(83)* are added, the result is:

$$P_1 + P_2 = \frac{\left[\dfrac{m_1}{(1 - B'^2_1)^{\frac{1}{2}}} + \dfrac{m_2}{(1 - B'^2_2)^{\frac{1}{2}}}\right] cB_3}{(1 - B^2_3)^{\frac{1}{2}}} \qquad (84)$$

Using the velocities of the primed reference frame allows P_1 and P_2 to have equations of like terms that can be added together directly. The spring mass property doesn't have any effect on the calculations. Equation *(84)* gives the correct momentum, and P_3 of *(80)* is not correct. The spring mass property will be included and the momentum from the unprimed reference frame will be recalculated.

The mass property of the potential energy of the spring at rest is needed. The coiled spring is at rest in the primed reference frame. As before, from the Law of Conservation of Energy, the potential energy of the spring will be equal to the kinetic energy of the two blocks before their collision.

$$m_p = \frac{\left[\dfrac{m_1 c^2}{(1 - B'^2_1)^{\frac{1}{2}}} - m_1 c^2\right] + \left[\dfrac{m_2 c^2}{(1 - B'^2_2)^{\frac{1}{2}}} - m_2 c^2\right]}{c^2} \qquad (85)$$

This is the mass property of the spring in the primed reference frame, where it is at rest. Note that the two blocks are also at

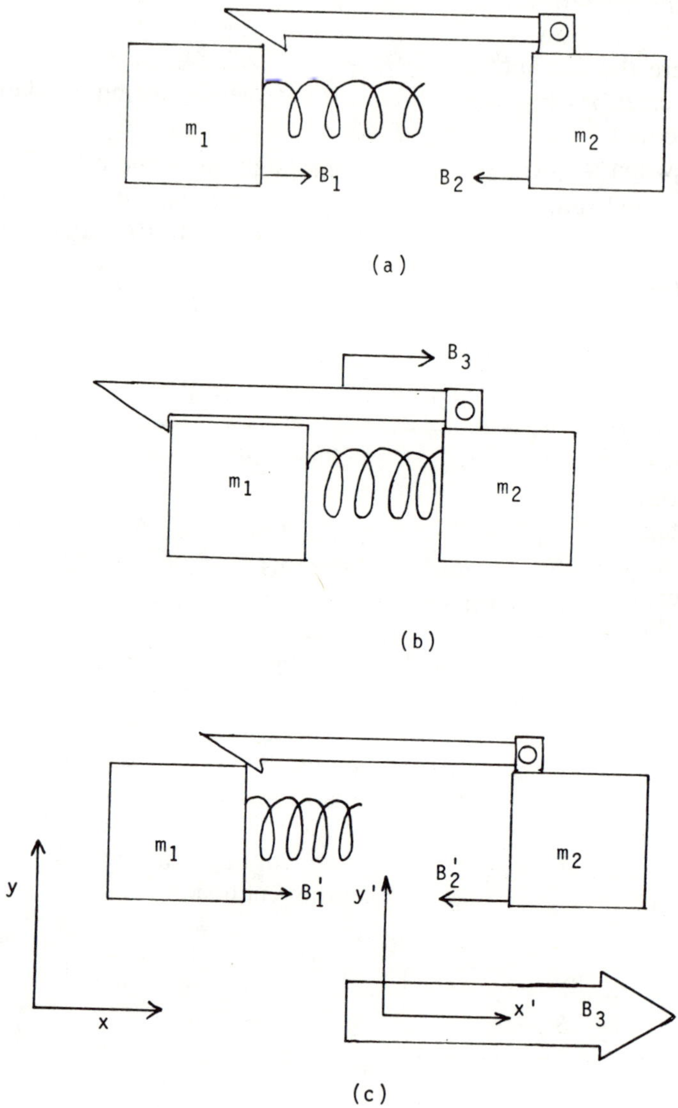

Figure 12

rest in the primed reference frame when they are locked together as in Figure 12b. The mass property of the spring can be used anywhere the rest masses of the two blocks are used in mass-related calculations. If the mass property of the spring is taken into account in the expression for the final momentum of the system when it is locked together, the result is:

$$P'_3 = \frac{(m_1 + m_2 + m_p)cB_3}{(1 - B_3^2)^{\frac{1}{2}}} = \frac{\left[\frac{m_1}{(1 - B'^2_1)^{\frac{1}{2}}} + \frac{m_2}{(1 - B'^2_2)^{\frac{1}{2}}}\right] cB_3}{(1 - B_3^2)^{\frac{1}{2}}} \tag{86}$$

Equations *(86)* and *(84)* are equal. $P_1 + P_2$ is equal to P'_3, and the momentum before the interaction with the spring is the same as the momentum after the interaction. The Law of Conservation of Momentum is satisfied when the mass property of the potential energy of the spring is considered.

This example has used the mechanism to latch the blocks together. If one considers any instantaneous situation, however, where the blocks are running into the spring, the Law of Conservation of Momentum should still be dependent upon the potential energy of the spring acquired to that point in time.

As one final example, consider the block of Figure 13. A spring inside the block acts on a piston as shown. The spring is

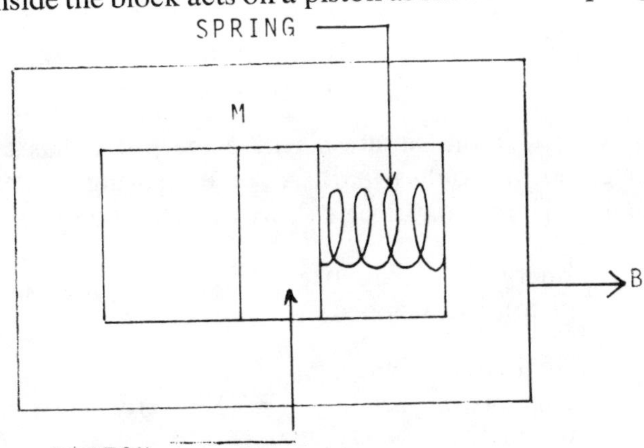

Figure 13

held in position by a mechanism which fixes the position of the piston relative to the object. This mechanism can release the piston when desired. The surfaces of the cylinder which the piston slides on when released exert a frictional force on the piston. The relationship between the frictional force, the mass of the piston, and the spring rate is such that, when the piston is released, it will be stopped by the friction just as the spring reaches its undeflected length, so that all of the spring potential energy is changed into heat energy. The mass of the block and piston is $m - m'$. The heat produced is the same amount produced if mass m' were turned into heat. The momentum of the block with the spring compressed is:

$E/c^2 = m_{sp}$ — mass property of spring potential energy, E-potential energy of spring

$$P = \frac{(m - m' + m_{sp})cB}{(1 - B^2)^{\frac{1}{2}}}$$ momentum of block with spring compressed

$E = m'c^2$ — spring potential energy and heat energy relationship

$m' = m_{sp}$

$$P = \frac{mcB}{(1 - B^2)^{\frac{1}{2}}}$$ momentum of block, spring compressed (87)

Consider the situation where the piston has traveled far enough so that only a part, A, of the spring potential energy remains. The piston kinetic energy is negligible.

Energy in spring after piston has stopped $= AE \qquad 0 \leq A \leq 1$

Heat energy generated after piston has stopped $= (1 - A)E$

The momentum of the block for any value of A is:

$m'_{sp} = AE/c^2 = Am_{sp}$ remaining spring mass property

$m_h = (1 - A)E/c^2 = (1 - A)m_{sp}$ mass property of heat generated thus far

$$P = \frac{(m - m' + m'_{sp} + m_h)cB}{(1 - B^2)^{\frac{1}{2}}}$$
$$= \frac{[m - m' + Am_{sp} + (1 - A)m_{sp}]cB}{(1 - B^2)^{\frac{1}{2}}}$$

$$P = \frac{mcB}{(1 - B^2)^{\frac{1}{2}}} \tag{89}$$

The momentum of the block is still conserved. This brief exercise is intended to show how momentum is constant even as energy within the system is changing from one form to another. The Law of Conservation of Energy and Law of Conservation of Momentum are in fact the same law with the concept of mass property.

Potential energy has mass property, just as heat, light, and kinetic energy. This will end this chapter on mass property. There are still other forms of energy to consider, but dealing with them should not be difficult. The basic concepts have been established and are all that are needed to explore the mass property of other energy forms. In fact, the examples presented could be expanded into many examples dealing with different facets of mass property without much difficulty. This chapter has not been intended to be a catalogue of proofs for every conceivable situation, but merely a demonstration of how one of the concepts of relativity can be studied. As an aid in dealing with other experiments, it should be remembered that any form of energy can be converted into heat, light, kinetic, or potential energy. That and the Law of Conservation of Momentum and Newton's Law are the basic tools for this investigation.

Also of importance in this discussion is the concept of total mass. This says basically that each form of energy of an object contributes to the momentum of the object. Heat energy is "extra mass" that must be accounted for in certain circumstances if the right results are to be achieved. When the practical side of trying to separate the mass of an object from the heat contained in the object is considered, a problem results. Theoretically, all objects would collapse if all of the heat is removed from them (the temperature of the object would be at absolute zero). Objects do not exist in an easily definable state unless they are at a temperature above absolute zero. For all practical purposes, an object and some of its heat energy cannot be separated, so defining mass of an object as the total mass is just another name for treating the mass of objects as it has always been done. The reason total mass has not been required until relativity is probably because the heat energy of an object is not large in comparison to the energy of the mass of the object.

The potential energy also should be considered as part of the total mass of an object. The light energy, while not really part of an object as light energy, can be part of a system of objects. The heat, potential, and light energy of a system comprise the basic total mass of the system. This total mass is in the rest reference frame of the system. The kinetic energy of the system expands the energy of the system at rest as a function of velocity, and heat and potential energy especially can influence what the kinetic energy and momentum of a system look like. In fact, the kinetic energy of a system can be regarded as part of the basic total mass in certain circumstances (such as the case of Figure 7). The concept of total mass basically says that notions of mass variability, such as is represented by *(38)* and *(61)*, are not valid. The material mass does not change functionally as *(61)* would indicate. What is happening is that, as an object gains speed, the additional mass property of the kinetic energy of the object will increase the momentum of the object similar to the increase in momentum that would occur if the mass of the object did vary according to *(61)*.

In summary, there are examples of experiments available that clearly violate Newton's Law or the Law of Conservation of Momentum unless certain forms of energy are considered to interact inertially as part of the system. This characteristic of energy is mass property. The concept of mass property is a form of the statements of the Theory of Relativity that matter and energy are different forms of the same thing.

Chapter Three

GRAVITY AND ENERGY

Gravity is a force that exists mutually between two masses. The magnitude of the force is related to the masses involved, and is explained by two theories. The General Theory of Relativity is the most sophisticated interpretation of the gravitational phenomenon. Gravity is, however, approximated by Newton's Inverse Square Law.

$$F = \frac{Gm_1m_2}{r^2} \tag{90}$$

F = mutual force between two masses
m_1 = mass value of one of the masses
m_2 = mass value of the other mass
r = distance between the masses
G = gravitational constant

In the preceding chapter, the concept of mass property of energy was examined regarding the inertial interactions of energy in a system of matter and energy. Energy reacts to Newton's Law as material objects do. Mass property relates the inertial interaction of energy in the same way as mass functions for material objects. In this chapter, gravity and its relevance to energy due to the energy mass property will be examined.

The gravitational force field surrounding a mass is a conservative force field. In the case of an object near the earth's surface, the gravitational potential energy of the object will increase if the object is moved farther from the earth. If an object is allowed to fall freely, the potential energy is converted into kinetic energy. Consider the object with mass m of Figure 14. The object is a distance R_1 from the earth, which has mass m_e. The object's mass m is small compared to the earth's, so that, in a free state of motion, the earth remains essentially stationary and the object will fall toward the earth under the gravitational attraction present. When it is time for this experiment to begin, the object begins to move to a distance R_2 from the earth. The earth is used here because it is a large familiar body. The earth is not a perfect sphere, nor is it perfectly homogeneous. These two factors will cause irregularities in the gravitational field. In addition, the rotation of the earth will add a centripetal acceleration factor to the gravitational acceleration, causing further deviations from the ideal gravitational field

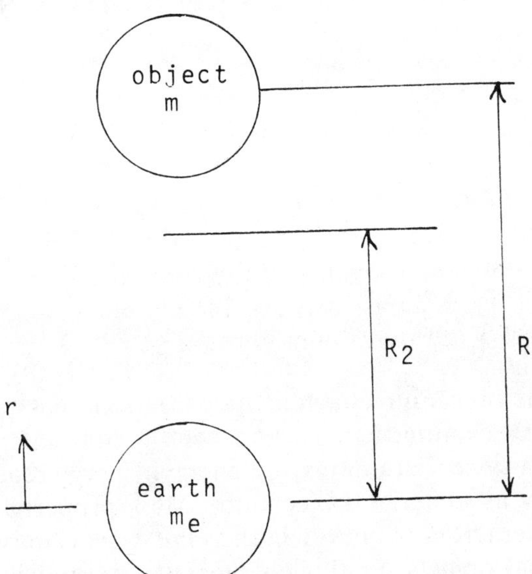

Figure 14

[which may be represented approximately by *(90)*]. All of the bodies used in this discussion will be assumed to be spherical, homogeneous, and nonrotating so that the gravitational field is the mathematical ideal. The mass of the bodies will be assumed to be concentrated at a point at the geometric center of the body. Under these circumstances, the change in potential energy of the object as it travels from R_1 to R_2 is:

$$V_o = -\int_{R_1}^{R_2} F_o \, dr \qquad F_o = -\frac{Gmm_e}{r^2}$$

$$V_o = -\int_{R_1}^{R_2} -\frac{Gmm_e}{r^2} \, dr$$

$$V_o = Gmm_e \left(\frac{1}{R_1} - \frac{1}{R_2}\right) \tag{91}$$

The potential energy change going from R_1 to R_2 is negative. The object is assumed to have more potential energy when it is at the distance farther from the earth. Going to the distance which is closer to the earth represents a loss in potential energy. The force on the object, F_o, is in the negative r direction.

The same experiment will be repeated with a slight modification. Suppose a portion of the object's mass, m', is converted into heat energy, and the heat is contained in the object because the exterior of the object is a layer of insulation. The mass of the object was m before part of it is converted to heat (so the previous exercise on the potential energy change applies to the object before the mass is converted to heat). It is desired to find out if the earth's gravity is exerting a force on the heat energy in the object. This is accomplished by supposing, first, that the heat energy is not affected by gravity. When part of the object's mass is converted into heat, this will mean that gravity will exert less of a force on the object. In this case, the potential energy change of the object as it goes from R_1 to R_2 will again be calculated.

$$F_1 = -\frac{Gm_e(m - m')}{r^2}$$

$$V_1 = Gm_e\,(m - m')\left(\frac{1}{R_1} - \frac{1}{R_2}\right)$$

$$F_1 \neq F_o$$
$$V_1 \neq V_o \qquad (92)$$

The force on the object, F_1, is less than it was before the mass was converted into heat, and the potential energy change, V_1, is not the same as V_o.

The system has not been changed except for the conversion of some mass to heat. By assuming that gravity does not affect the heat energy, this means that the energy of the system is not the same as it was before the mass conversion. The change in the potential energy difference means that the Law of Conservation of Energy has not been observed. The Law of Conservation of Energy states that the energy of a system cannot be created nor destroyed, only changed in form. The loss of some of the gravitational potential energy given by *(92)* when compared to *(91)* is a loss of some of the energy of the system.

That the Law of Conservation of Energy is contradicted if heat energy is not attracted by gravity can be shown by expanding on the previous process and using the potential energies of *(91)* and *(92)*. Suppose an electric motor is used to change the position of the object, as is illustrated in Figure 15. As the object descends to the earth, the electric motor acts as a generator to charge the battery that is connected to it, and draws this energy from the change in gravitational potential energy. As the object ascends from the earth, the motor expends some of the work stored in the battery. The wires and battery will be assumed to have no resistance, and the motor will be assumed to be 100% efficient in all its operations.

Using this apparatus, the total system energy will be calculated. The energy represented by the earth's mass is $E_1 = m_e c^2$; the energy represented by the object's mass is $E_2 = mc^2$; and, if R_2 is such that the object is resting upon the earth at R_2, the maximum potential energy available is V_o. The process is that the object goes from R_1 to R_2, where the object rests on the earth, converting the potential energy into additional energy of the earth (the battery is assumed to be part of the earth). The object and the earth then represent energy $E_t = (m_e + m)\,c^2 + V_o$. This is the total energy of the system.

Now, suppose that the object is raised back to R_1, withdrawing V_o energy from the battery and restoring the system to its initial state. Mass m' in the object is converted into heat energy $E_3 = m'c^2$. This energy is not affected by gravity by assumption. The object is then lowered back down to R_2 to rest again upon the earth's surface. The energy received by the battery is V_1 and the energy represented by the remaining portion of the object's matter is $E_4 = (m - m')c^2$. The total energy of the system is now $E'_t = (m_e + m' + m - m')c^2 + V_1 = (m_e + m)c^2 + V_1$. E'_t is less than E_t, however. The energy of the system is now less than it was initially, so some energy has

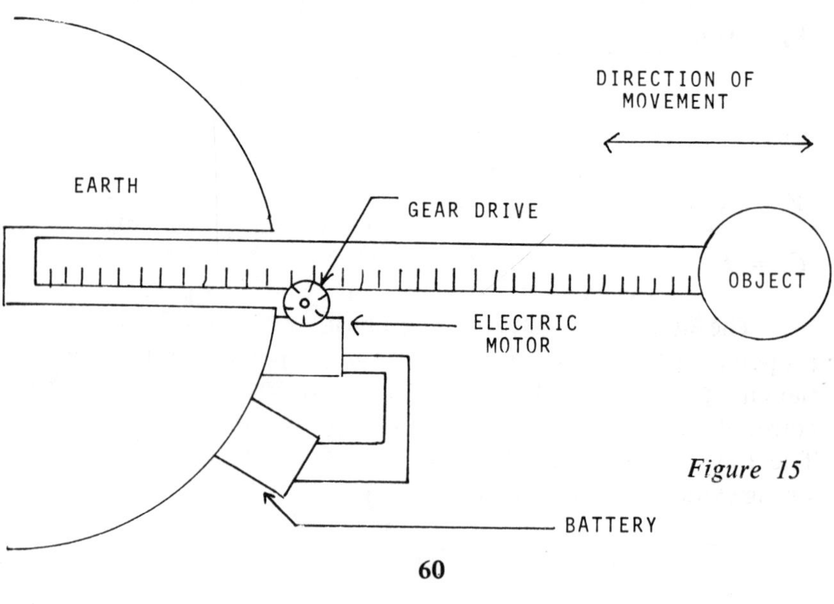

Figure 15

been destroyed. This result is contrary to the Law of Conservation of Energy. The assumption that gravity does not affect the heat energy must be invalid.

The calculation of the potential energy change of the object as it goes from R_1 to R_2 will again be done, this time assuming that gravity affects the heat energy within the object. The mass property of the heat energy of the object will be assumed to have the same function as mass does for matter in determining the effect of gravity on the energy. Mass property will be defined as it was in the previous chapter.

$E = m'c^2$ — heat energy of converted mass m'

$m_h = E/c^2$ — definition of heat mass property

$m_h = m'$

$V_2 = Gm_e(m - m')\left(\frac{1}{R_1} - \frac{1}{R_2}\right) + Gm_e m_h \left(\frac{1}{R_1} - \frac{1}{R_2}\right)$ —
potential energy change of mass of object's matter, and potential energy change of heat energy in object

$$V_2 = Gm_e m\left(\frac{1}{R_1} - \frac{1}{R_2}\right)$$

$F_2 = -\frac{Gm_e m}{r^2}$ — force needed to produce V_2

$$V_o = V_2$$

$$F_o = F_2 \qquad (93)$$

The assumption that gravity affects mass and heat mass property in the same manner results in the gravitational potential energy change of the object as it goes from R_1, to R_2 being constant throughout the process of mass conversion to heat. The Law of Conservation of Energy is satisfied, and that is strong evidence in favor of this assumption. Note that no specific values have been given to R_1 or R_2. For either case of the object with or without some mass converted to energy, the effects of the gravitational force field on the object is the same for any set of R_1 and R_2. The potential energy change must be functionally identical for both cases. Since the force necessary to produce the potential energy change is a singular mathematical function of the potential energy change, the force on the object in either case must be the same, as is indicated in the respective equations. Note also that if the object has any significant velocity as it goes from R_1 to R_2, it will have kinetic energy in that direction. In order that each experiment be as simple as possible, this kinetic energy will be avoided in all of the experiments of this chapter by assuming the object travels "slowly" from R_1 to R_2.

The same method can be used to show that gravity acts on the mass property of potential energy. Consider the object of Figure 16. Initially, the object is held at a distance R_1 from the earth similar to the situation of Figure 14. In the situation of Figure 16, a spring inside the object is acting on a piston as

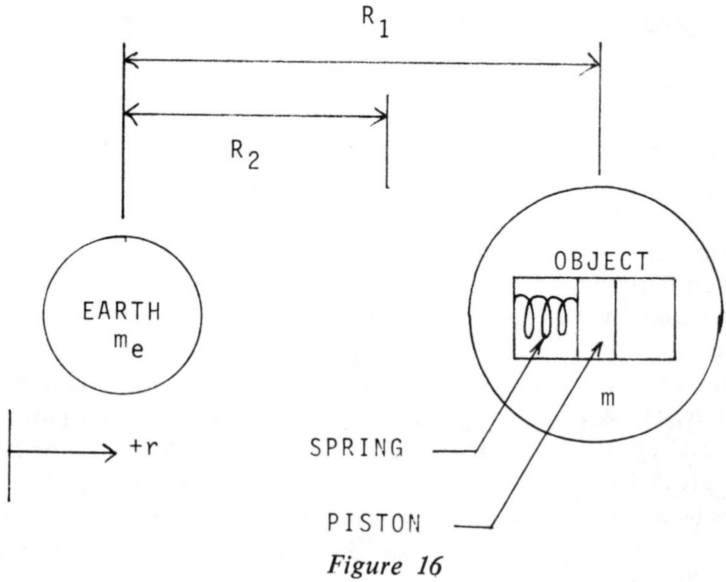

Figure 16

shown. The spring is held in position before the experiment starts by some mechanism which can be released as desired. The piston slides on surfaces of the cylinder in which it is located when the spring is released, these surfaces exerting a frictional force on the piston. The relationship between this frictional force, the mass of the piston, and the spring rate is such that when the spring is released and pushes the piston in the cylinder, the piston will come to a stop just as the spring reaches its free length (undeflected length). This means that all of the energy stored in the spring is changed into heat energy, generated by the friction. The spring and piston mass are part of the object mass which is $m-m'$. The amount of heat released into the object by the frictional process is equal to the amount of heat produced in the previous experiment when mass m' was converted into heat. This means the potential energy stored in the spring is also equal to the heat energy produced by conversion of m' in the previous experiment. As before, the exterior of the object is insulation which contains all of the heat within the object. The mass of the object plus the mass property of the spring's potential energy or the heat energy give a total mass of the object of m.

Two experiments can be done with this apparatus. The first is to allow the object to go from R_1 to R_2 with the spring in the stressed position; the second is to allow the spring to move the piston, producing heat energy, and then to allow the object to go from R_1 to R_2. In the case where the object contains the heat energy, the experiment has been set up so that the situation is essentially the same as the experiment of Figure 14 when that object contains heat energy. The amount of heat energy present is the same, and the mass of the object without the heat energy is the same. The change in potential energy as the object goes from R_1 to R_2 for the situation of Figure 16 after the spring potential energy has been converted into heat energy is therefore given by *(93)*. It has been established that gravity influences heat energy.

It is desired to prove that the spring potential energy is also affected by gravity. The initial assumption in this proof will be that gravity does not affect the spring's potential energy. This will mean that the force on the object with the spring stressed will be less than the force with the spring's energy converted into heat energy. The calculation for the gravitational potential energy change of the object as it goes from R_1 to R_2 in this situation is:

$$F_3 = -\frac{Gm_e(m - m')}{r^2}$$

$$V_3 = Gm_e(m - m',\left(\frac{1}{R_1} - \frac{1}{R_2}\right)$$

$$F_3 \neq F_2 = F_o$$
$$V_3 \neq V_2 = V_o \qquad (94)$$

With the assumption that gravity does not affect the spring's potential energy, the gravitational potential energy change as the object goes from R_1 to R_2 is not the same as it would be if the spring's energy were converted to heat. Since it is optional at the start of the experiment whether the spring's potential energy or heat energy will be present during the experiment, doing the experiment with the spring's energy would result in less energy being present in the system with the object at R_2. This represents a loss of energy from the system without any energy crossing the system boundaries. If the experiment is done with the spring stressed, and at R_2 the spring is released, creating heat energy, the final state of the object at R_2 would be the same as if the experiment were done with the heat energy. The initial and final states of the system are the same, though differing amounts of energy are produced, depending on which experiment is performed. This loss of energy is in conflict with the Law of Conservation of Energy.

The experiment will be repeated, but the assumption this time will be that gravity affects the mass property of the spring's potential energy in the same way as it affects heat energy. The mass property of the spring potential energy will be m_s.

$E = m'c^2$ amount of heat energy in object (experiment of Figure 14 or 16)

$m_s = E_s/c^2$ definition of spring mass property $E_s =$ potential energy stored in spring

$$E_s = E$$

$$m_s = m'$$

$$V_4 = Gm_e(m - m')\left(\frac{1}{R_1} - \frac{1}{R_2}\right) + Gm_e m_s\left(\frac{1}{R_1} - \frac{1}{R_2}\right)$$

potential energy change of mass of object and potential energy change of spring potential energy.

$$V_4 = Gm_e m\left(\frac{1}{R_1} - \frac{1}{R_2}\right) \qquad F_4 = -\frac{Gm_e m}{r^2}$$

$$V_4 = V_2 \qquad F_4 = F_2 \qquad V_4 = V_o \qquad F_4 = F_o \tag{95}$$

The change in gravitational potential energy going from R_1 to R_2 is the same for the object with the stressed spring as it was for the object with the heat. The assumption that gravity affects the spring's potential energy is consistent with the Law of Conservation of Energy. The force function for the gravitational pull is the same for the spring energy and the heat energy if the potential energy change is the same for both cases. The result is that gravity exerts the same influence over the mass property of the spring's potential energy as it does on the mass of the matter in the object.

As one final example of the calculation technique used in determining the gravity-energy relationship, the gravitational effect on kinetic energy will be determined. This experiment involves the situation of Figure 17. Two identical blocks of mass m_1 are shown inside the object, each block with a velocity opposite in direction but equal in magnitude to the others. The blocks are sliding on frictionless surfaces in straight tracks. The paths of the blocks are such that no net momentum exists in the

x direction and the object has no net angular momentum about any axis, the momentum of each block cancelling out the others in a calculation of the overall object momentum. Four identical springs are arranged as shown, one at each end of each track. The length of each track is the same, and the positions of the blocks is such that each block reaches the end of its track, and consequently runs into its respective spring, at the same time. The purpose of the springs is to rebound each block back across the track in the direction opposite to that which it came from. Note the orientation of the track lengths with respect to the earth. Because the blocks run into the springs simultaneously, the net force on the object produced by the compressing of the springs when the blocks run into them is zero. The intent of this arrangement is to have the blocks continually running back and forth in the tracks, rebounding off the springs indefinitely, the motion of each block appearing to be the mirror image of the other. The object will contain some mechanism for grabbing each block simultaneously at the end of the tracks when their

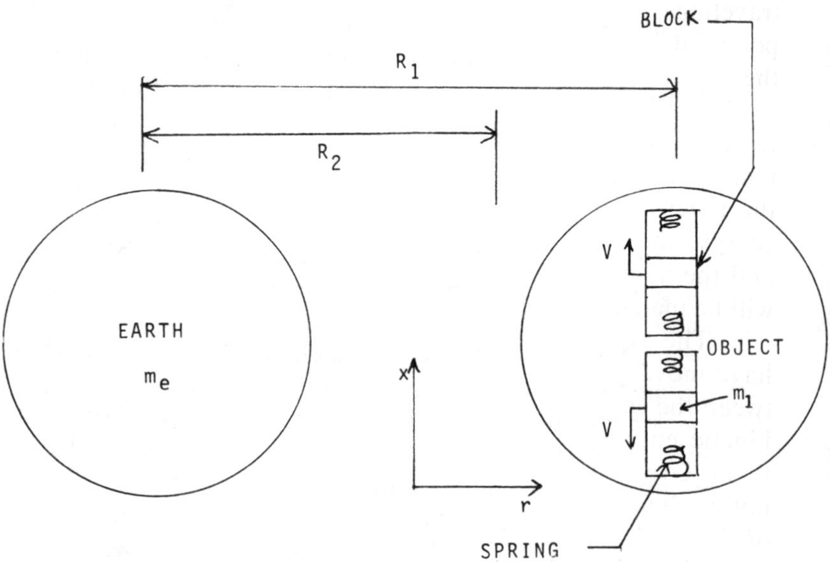

Figure 17

velocity is zero. This is the point of maxium compression of the springs involved. This mechanism can be activated when desired. When this mechanism has been activated, the blocks are stationary and the springs are in the position of maximum potential energy. The mass of the object at this point is the mass of the matter of the part of the object which makes up the grabbing mechanism, tracks, and other material which is not active, the mass of the two blocks, and the mass of the springs and the springs' potential energy. This total object mass is m. The mass property of the potential energy of two compressed springs will be equal to m'.

With the blocks oscillating as described, there will be a continual flow of energy in the object from kinetic energy of the blocks to potential energy of the springs and back again. It has been shown that gravity affects the potential energy of the springs. The object is again at a distance R_1 from the earth at the start of the experiment. If the blocks were locked in a stationary positon by the grabbing mechanism, the object, upon traveling from R_1 to R_2, would experience the gravitational potential energy change given by *(95)* or *(91)*. The springs of the object of Figure 17 would contain the same potential energy as the spring of Figure 16. If the grabbing mechanism is reeased, allowing the blocks to move across the tracks, the potential energy of the springs is changed into kinetic energy of the blocks. To show that gravity has an effect on the kinetic energy, the preceding calculation procedure will be repeated, and the assumption that gravity does not affect kinetic energy will be made.

The experiment begins with the object at R_1. Both blocks have moved to that section of their tracks where they are between the springs (not touching any springs). The blocks have kinetic energy equal to the spring's potential energy given to them. Gravity still has an effect on the mass of the blocks but not on their kinetic energy. The relationship between the length of the track, velocity of the blocks, and R_1 and R_2 is such that the object reaches R_2 before the blocks run into any springs. With the blocks moving in the tracks, the mass of the object which gravity acts on is reduced by the amount of the mass property of the spring's potential energy.

m_o = mass of grabbing mechanism, tracks, and object proper

m_1 = rest mass of one block

m_s = mass property of springs' potential energy at maximum spring deflection

$m = m_o + 2m_1 + 2m_s$ = total mass of object with blocks stationary and springs fully compressed

$m_r = m_o + 2m_1$ = total mass of object with blocks moving between springs (gravity does not affect blocks' kinetic energy) and springs in the undeflected state

$V_5 = Gm_e(m - 2m_s)\left(\frac{1}{R_1} - \frac{1}{R_2}\right)$ = change in potential energy with blocks moving in their tracks

$F_5 = -\frac{Gm_e(m - 2m_s)}{r^2}$ = force needed to produce V_5

$$V_5 \neq V_o \qquad F_5 \neq F_o \tag{96}$$

The total mass of the object that gravity has an effect on with the springs fully compressed is m, the same mass as the object had in Figure 14. The potential energy change of the situation of Figure 17 can be compared to that of Figure 14. The total mass of the object that gravity has an effect on with the blocks moving in their tracks is m_r which is less than m because of the assumption that gravity does not affect the kinetic energy of the moving blocks. This kinetic energy was the springs' potential energy before the springs moved the blocks into the track space between them. The potential energy change of the object in this case is V_5 and is less than that of *(91)*. The potential energy change of the object should be the same as *(91)* and *(95)* because the mass property of the fully compressed springs of this case is the same value as the mass property of the fully compressed spring of Figure 16, to which *(95)* applies. Therefore, by assuming that gravity does not affect the kinetic energy of the blocks, there is a contradiction to the Law of Conservation of Energy similar to the previous contradictions noted. If the object is held at R_2 and the blocks are allowed to rebound back across the tracks to their original positions at the start of the experiment, the initial and final situations of the object will be the same as they would be if the object had gone from R_1 to R_2 with the blocks fixed and the springs fully compressed [a situation described by *(95)*]. The force on the object in *(96)* which would be necessary to produce V_5 reflects that gravity does not have an effect on the blocks' kinetic energy.

Having arrived at a contradiction to the Law of Conservation of Energy when it is assumed that gravity does not affect kinetic energy, the assumption will now be made that gravity does influence kinetic energy. The kinetic energy of the blocks is equal to the potential energy stored in the springs, so the mass property of this kinetic energy is equal to the mass property of the spring potential energy.

m_k = mass property of one block's kinetic energy

$$m_k = m_s$$

$m'_r = m_o + 2m_1 + 2m_k$ = total mass of object with blocks moving between springs; springs in undeflected state

$m = m_o + 2m_1 + 2m_s$ = total mass of object with blocks stationary and springs fully compressed

$$m = m'_r$$

$$V_6 = Gm_e m\left(\frac{1}{R_1} - \frac{1}{R_2}\right) = \text{change in potential energy with blocks moving in tracks}$$

$$F_6 = -\frac{Gm_e m}{r^2} = \text{force needed to produce } V_6$$

$$V_6 = V_o \qquad F_6 = F_o \tag{97}$$

The total mass of the object that gravity influences is a constant throughout the process where potential energy changes to kinetic energy if the assumption that gravity influences kinetic energy is made. The gravitational potential energy change of *(97)* is the same as *(91) or (95)*. The Law of Conservation of Energy is satisfied. The force needed to produce V_6 is the same as that needed to produce V_o, reflecting gravitational effect on kinetic energy mass property.

The effect of gravity of mass (matter) on energy has been established. Since the gravitational attraction between two bodies is a mutual attraction, the possibility that energy exerts a gravitational force of its own on matter will be investigated. Through gravity, the matter of the earth attracts the energy in the object, but that does not mean the energy of the body is attracting the matter of the earth. The gravitational attraction that energy has for matter will be investigated using the situation of Figure 14. In *(91)*, the potential energy change the object experiences as it travels from R_1 to R_2 is given if both the object and the earth are all matter. As before, this experiment will be repeated with a modification. Suppose a portion of the earth's mass, m'_e, is converted into heat energy, and the heat is contained in the earth because the exterior of the earth is a layer of insulation. The mass of the earth was m_e before part of it was converted to heat, so the exercise of *(91)* gives the potential energy change that is expected. The mass of the object is m, and no part of the object is converted into heat. To find if the energy of the earth gravitationally attracts the matter of the object, it will first be assumed that energy exerts no gravitational influence over matter. The change in gravitational potential energy is recalculated with this assumption in force.

$$F_7 = -\frac{Gm(m_e - m'_e)}{r^2} = \text{force on object}$$

$$V_7 = Gm(m_e - m'_e)\left(\frac{1}{R_1} - \frac{1}{R_2}\right)$$

$$F_7 \neq F_o \qquad V_7 \neq V_o \tag{98}$$

The assumption that energy exerts no gravitational influence over matter gives a reduced gravitational force on the object, F_7, as compared to F_o. Consequently, the gravitational potential energy change, V_7, is not the same as the original, V_o. As before, because it is optional whether or not the part of the earth is converted into energy at the beginning of the experiment, the potential energy change V_7 represents a loss of energy of the system while no energy crosses the system boundaries. This is in contradiction with the Law of Conservation of Energy. This may be interpreted to mean the assumption that energy does not gravitationally influence matter is invalid. The experiment will be recalculated, assuming that energy does have a gravitational influence over matter. It will be assumed that the gravitational force involved is related to the mass property of the heat energy in the same way that gravity relates to the mass of material.

$$E = m'_e c^2 = \text{heat energy of converted mass } m'_e$$

$$m_{eh} = E/c^2 = \text{definition of mass property}$$

$$m_{eh} = m'_e$$

$$V_8 = Gm(m_e - m'_e)\left(\frac{1}{R_1} - \frac{1}{R_2}\right) + Gmm_{eh}\left(\frac{1}{R_1} - \frac{1}{R_2}\right)$$

= potential energy change due to gravitational force of earth's matter and of earth's heat energy

$$V_8 = Gmm_e\left(\frac{1}{R_1} - \frac{1}{R_2}\right)$$

$$F_8 = -\frac{Gmm_e}{r^2} = \text{force needed to produce } V_8$$

$$V_8 = V_o \qquad F_8 = F_o \tag{99}$$

The assumption that energy exerts a gravitational influence over matter has resulted in a gravitational potential energy change that is consistant with *(91)*. This is in agreement with the Law of Conservation of Energy. The force needed to produce V_8 is the same magnitude as that needed to produce V_o, which indicates that the gravitational property of the mass property of heat energy is of the same form as the gravitational property of the mass of matter. There is not much difference between *(99)* and *(93)*. The gravitational properties of potential and kinetic energy can be established using the same variation of the heat energy principle demonstrated in *(98)* and *(99)*.

In addition to the discussions proving the gravitational attractions of matter for energy and energy for matter, a further modification of the example of Figure 14 will show the gravitational attraction of heat energy for heat energy. The experiment will be designed so that *(91)* again serves as the potential energy change to which the other potential energies of the modified experiment will be compared. Suppose a portion of the object's mass, m', is converted into heat, and a portion of the earth's mass, m'_e, is converted into heat simultaneously. The heat is contained within both bodies as the exterior of each body is a layer of insulation. The initial mass of the object before the heat conversion is m and the initial mass of the earth before the heat conversion is m_e. The proof that energy exerts a gravitational attraction over·energy will be done by first assuming that energy does not exert a gravitational influence over energy. It has already been established that the matter of the earth attracts the matter in the object, that the matter of the earth attracts the energy in the object, and that the energy of the earth attracts the matter in the object. The potential energy and force calculation is;

$$E_e = m'_e c^2 = \text{heat energy of earth produced by } m'_e$$

$$m_{eh} = E_e/c^2 = \text{definition of mass property}$$

$$m_{eh} = m'_e$$

$$E = m'c^2 = \text{heat energy of object produced by } m'$$

$$m_h = E/c^2 = \text{definition of mass property}$$

$$m_h = m'$$

$$V_9 = G(m - m')(m_e - m'_e)\left(\frac{1}{R_1} - \frac{1}{R_2}\right) + Gm_h(m_e - m'_e)\left(\frac{1}{R_1} - \frac{1}{R_2}\right) + G(m - m')m_{eh}\left(\frac{1}{R_1} - \frac{1}{R_2}\right)$$

= potential energy change due to mass of earth attracting mass of object, mass of earth attracting energy of object, and energy of earth attracting mass of object

$$V_9 = G(mm_e - m'm'_e)\left(\frac{1}{R_1} - \frac{1}{R_2}\right)$$

$$F_9 = -\frac{G(mm_e - m'm'_e)}{r^2} = \text{force needed to produce } V_9$$

$$V_9 \neq V_o \qquad F_9 \neq F_o \qquad (100)$$

The gravitational force on the object F_9 reflects that the energy of the object is not attracted by the energy of the earth as was assumed. The potential energy change as the object goes from R_1 to R_2 is not the same as V_o, which is a contradiction to the

Law of Conservation of Energy. In order to obtain a calculation that is in agreement with the Law of Conservation of Energy, the assumption will be made that the energy of the earth attracts the energy of the object.

$$V_{10} = (Gmm_e - m'm'_e)\left(\frac{1}{R_1} - \frac{1}{R_2}\right) + Gm_{eh}m_h\left(\frac{1}{R_1} - \frac{1}{R_2}\right)$$

= potential energy change of object due to mass of earth attracting mass of object, mass of earth attracting energy of object, and energy of earth attracting mass of object (which is V_9) and due to the energy of the earth attracting the energy of the object

$$V_{10} = Gmm_e\left(\frac{1}{R_1} - \frac{1}{R_2}\right)$$

$$F_{10} = -\frac{Gmm_e}{r^2} = \text{force needed to produce } V_{10}$$

$$F_{10} = F_o \qquad V_{10} = V_o \tag{101}$$

As before, the gravitational potential energy change of the object as it goes from R_1 to R_2 would be reduced unless a gravitational attraction exists between the heat energy of the earth and the heat energy in the object. This assumption satisfies the Law of Conservation of Energy. The force needed to produce the gravitational potential energy change is the same magnitude force that is required if no heat energy is present. The discussions on the gravitational attraction of heat energy and matter, and the gravitational attraction of heat energy for heat energy can be extended to other forms of energy in the same manner as was presented for potential energy and kinetic energy and the effect of gravity on them. Gravity is a phenomenon which affects matter and energy in the same ways. The mass property of energy is equivalent to the mass of matter in the gravitational interactions involving energy.

A catalogue of proofs for the gravitational attraction of every type of energy will not be presented here. The technique for determining that gravity does have an effect on energy has been demonstrated. Showing gravitational effects on other types of energy will involve this technique with minor modifications depending on the energy type under consideration. The Law of Conservation of Energy is applied to material objects and energy as they undergo a change in gravitational potential energy. For the change in gravitational potential energy function, the force of gravity as a function of distance and total mass is used, based on the assumption of existence of a gravitational influence on the energy in question.

There has also been some discussion on the gravitational attraction that energy has for material and energy. The technique in these discussions is the same as that for showing the effects of gravity on energy. The behavior of energy in gravitational fields is one of the revelations predicted (and later experimentally verified) by the Theory of Relativity. The method used in relativity is vastly different from the one presented here.

Chapter Four

THE LAW OF CONSERVATION OF MOMENTUM AND THE LAW OF CONSERVATION OF ENERGY

Thus far, energy and momentum have been found to be conserved in every experiment ever designed to measure them. No deduction (other than this statement and "common sense") exists to prove these quantities are conserved in all situations. The experimental evidence has been so conclusive that it has been assumed these conservation laws must apply as fundamental laws of the universe. A fundamental law here refers to a law upon which other statements are based, which subtly defines it as being not provable by deduction from other statements. The only proof or disproof of a fundamental law lies with specific examples.

The Law of Conservation of Energy states that energy can neither be created nor destroyed within a system, only changed in form. The Law of Conservation of Momentum states that the total vector momentum of the system is a constant in the absence of forces outside the system, regardless of the events taking place in the system. In relativity theory, one form of these two laws is *(35)*, which combines them into a single mathematical statement. These laws will be examined in this chapter in light of the discussions of the previous chapters. Note that these laws are the building blocks of chapter two and chapter three.

First, the Law of Conservation of Momentum will be considered. In chapter two, using this law (and Newton's Law, which is essentially another form of this law), it was shown that energy reacts to forces as matter does and that energy possesses momentum as matter does. The discussion surrounding Figures 9 through 11 is of particular interest if the following discussion is to be understood. The experiment is an analysis of the mass property of potential energy, but has some value beyond that discussion. Note that when the m_1 blocks are moving as previously described, it is impossible for an observer on m_3 to tell when the blocks are moving or when they are stationary (and the springs compressed) by measuring any forces he might apply to m_2. The total mass of the blocks, springs, and framework is a constant with time. When the blocks are moving, their kinetic energy contributes to the total mass, and when the blocks are stationary, the springs' potential energy contributes to the total mass. One description of the situation might be that energy flows from the blocks to the springs and back again as the blocks run into the springs.

The machine of Figure 9b will be modified, and the system momentum again discussed. The modification described will be aimed not only at having the energy flowing between the blocks and springs, as described above, but also to have the energy flow from m_2 to m_3 and back again. The parts of the modified machine will be presented separately and the machine gradually built up.

The two m_1 blocks are still a part of this machine. They slide in straight, frictionless tracks, and have velocities equal in magnitude but opposite in direction at all times. The paths of the blocks are on the same straight line so that no net angular momentum is presented to m_2 at any time. Each track is the same length so each block travels the same distance. The blocks move, as before, as if each were the mirror image of the other. The total momentum of both blocks will always be zero in the direction of the tracks relative to m_2.

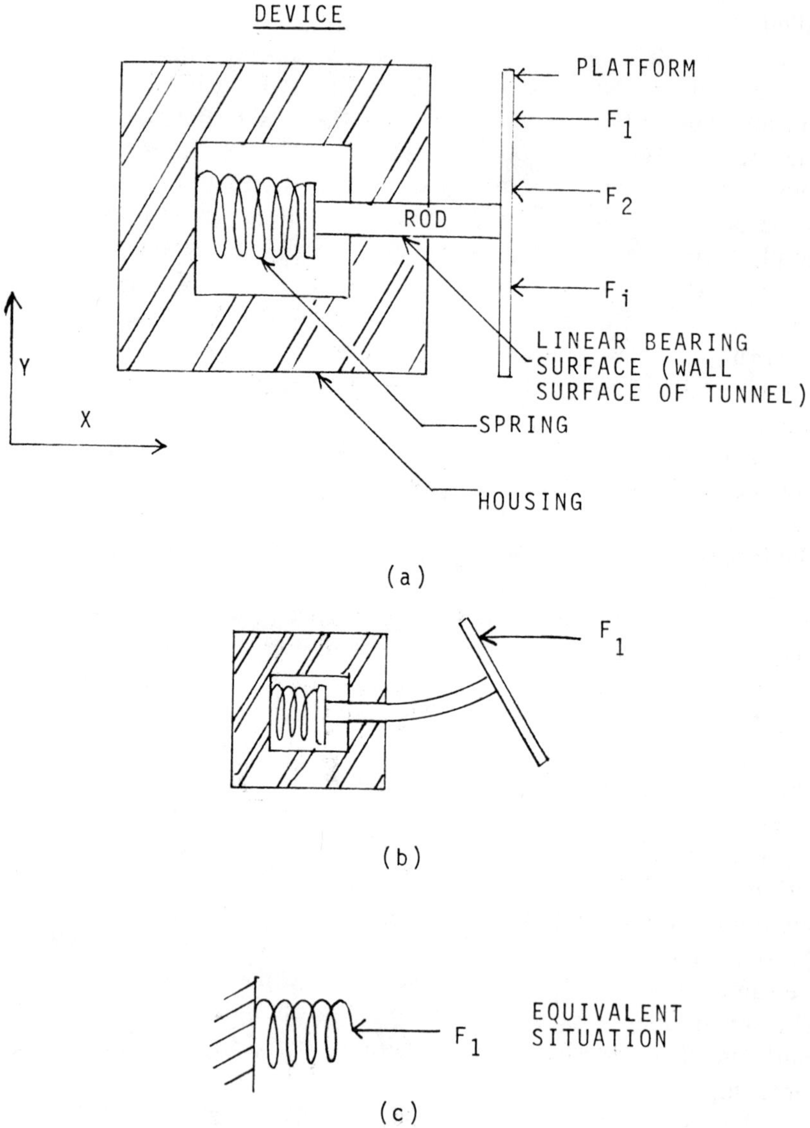
DEVICE
PLATFORM
F1
F2
Fi
ROD
LINEAR BEARING SURFACE (WALL SURFACE OF TUNNEL)
SPRING
HOUSING
Y
X
(a)
F1
(b)
F1
EQUIVALENT SITUATION
(c)

Figure 18

Now consider the device of Figure 18a. A solid object called the housing is drawn in cross section and has a cavity within it. A spring is mounted in this cavity that exerts a force on one end of a straight rod passing through a tunnel in the housing. The straight tunnel connects the cavity to the area outside of the housing. The rod goes completely through the housing to the area outside where it has a platform connected to the other end. All sliding surfaces, including the surface of the platform, are frictionless. The tunnel is a frictionless linear bearing surface which allows the rod to move in the x direction only (the rod cannot rotate). All parts are inflexible except the spring. The undeflected spring length is such that the spring may be undeflected and still fit within the cavity with room to spare.

The purpose of this device is to allow forces which are directed in the x direction to work on the spring, even though the force is not applied on a line concentric with the spring axis. The forces used in this example are labeled F_1, F_2, and F_i. The force F_1, for example, is applied at on end of the platform. If the parts were not frictionless and inflexible, the device would tend to distort and bind up, as is represented in Figure 18b. But with the assumptions made here, F_1 may be applied at the position shown and the reaction of the spring is the same as if F_1 were applied directly to the spring (see Figure 18c). The force F_2, which is of equal magnitude but in a different position, experiences the same spring reaction as F_1. Any force F_i, located at any position on the platform, will produce the same spring reaction with the device shown. Also, a force which is applied at one point on the platform and moved laterally by sliding across the frictionless platform surface to another point will produce the same spring reaction as a force which does not move at all. The spring does not know where the forces are applied on the platform. The spring can only react to the magnitude of the force impressed.

The device and the m_1 blocks are combined into the machine of Figure 19, shown in cross section to expose the parts of interest. The device is shown in four places, one device at each end of each track. The springs are in the undeflected state. The mass of the platforms and springs will be considered too small to be significant. The m_1 blocks are moving with speed $B = v/c$ relative to m_2. The m_1 blocks slide in straight, frictionless tracks in m_2. The straight tracks are held in a framework, m_2, which is of a slightly different shape than that of Figure 9b. The framework has a pole section which passes through the center of mass of m_2 and m_3. As before, m_3 is a large mass, but this time m_3 also serves as mounting points for the four devices. The pole slides on frictionless surfaces inside m_3 that are closely toleranced so that the relative motion of m_2 and m_3 is in the y direction only. There is a mechanism inside m_3 that can give m_2 and m_3 any desired relative motion by exerting a force on the pole. There is also a mechanism in m_2 that can grab the m_1 blocks simultaneously at either end of the tracks at the instant when their velocity relative to the framework is zero. As before, this is the situation where the springs have maximum deflections and maximum potential energy.

There is a dimension D shown on Figure 19. This is the clearance between the end of the track and platform with the spring in the undeflected state. The dimension d on Figure 19 represents the distance moved by the platform when the spring is at its maximum compression. The length of the block, L, is also shown. These dimensions are typical at all four platform locations. The condition that

$$D + d \ll L \tag{102}$$

must be specified so that the moving block is stopped by the platform and spring at the end of the track before the block slides out of the track completely. The track may be assumed to hold the blocks (due to this assumption) so that, at maximum spring deflection and with the grabbing mechanism holding the blocks stationary, a relative motion can exist between m_2 and m_3 without upsetting or tipping m_1 in the track. The m_1 block merely slides along the frictionless platform surface with the

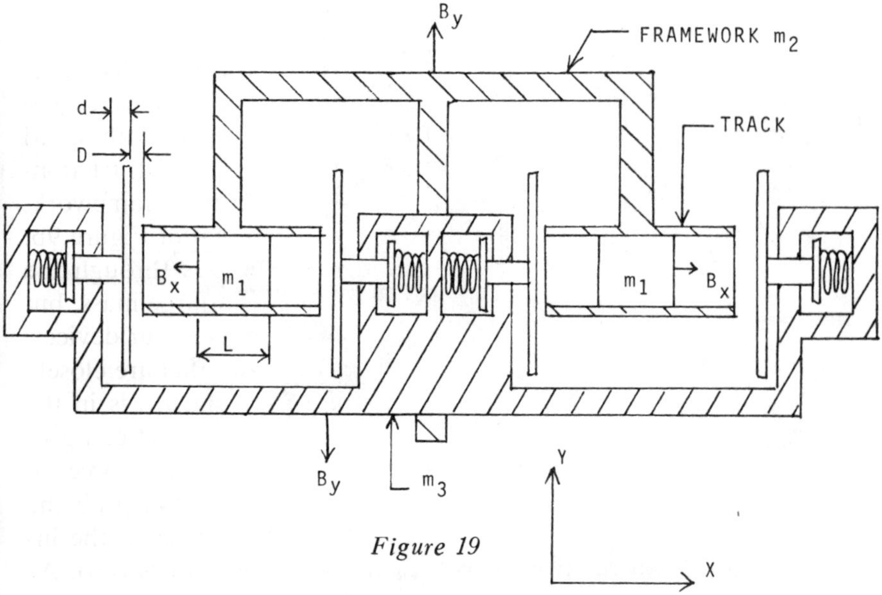

Figure 19

relative motion of m_2 and m_3. The m_1 blocks, if given a velocity, would continue to rebound back and forth across the tracks indefinitely, no matter what the motion of m_2 and m_3 (within the limits imposed by the dimensions of the platform).

The construction of this machine is, actually, similar to the machine of Figure 9b, with one difference. The springs are now part of m_3 instead of being part of m_2 as they were in Figure 9b. Consequently, the mass property of the springs' potential energy will now be part of the total mass of m_3 (where before it was a part of the total mass of m_2). The device is needed so that the springs may act on m_1 at any relative position or velocity of m_2 and m_3. The device was not needed in Figure 9b because making the spring part of m_2 was sufficient to put the potential energy mass property into the total mass of m_2.

The machine is to go through a process and the initial and final momentums of the machine will be compared. The initial stage of the process is shown in Figure 20. Both m_1 blocks are locked in a position by the grabbing mechanism where they have compressed two springs to their maximum deflection. The

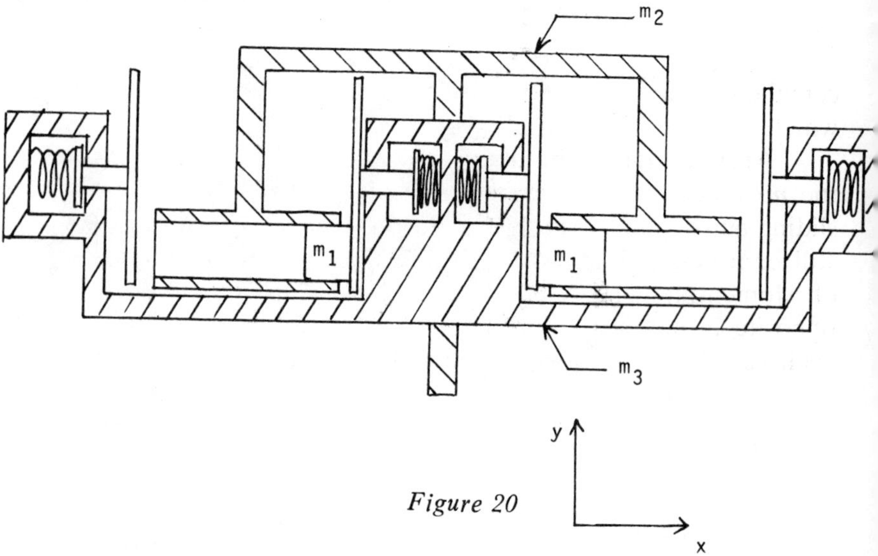

Figure 20

position of m_2 relative to m_3 is such that the m_1 blocks are positioned at the ends of the platforms they are touching. There is no relative m_2–m_3 motion and the machine is initially stationary with respect to the reference frame, so the initial momentum of the machine is zero. The mass property of each deflected spring's potential energy is m_s. The other springs are in the undeflected state, have no potential energy, and do not contribute to the total mass of m_3. For convenience, it will be assumed that:

$$m_3 + 2m_s = m_2 + 2m_1 \tag{103}$$

At this point, the mechanism which controls the relative motion between m_2 and m_3 gives each an impulse, resulting in a relative m_2–m_3 velocity. This velocity means that the parts of the machine have kinetic energy. This energy must have come from m_3, assuming the device responsible for the relative motion is located in m_3. The total mass of m_3 is diminished by the loss of this energy, but it will be assumed the velocities of m_2 and m_3 are small enough so that the loss of mass by m_3 is negligible. Even though the velocities are now assumed small, the relativistic momentums of the machine parts will be used in the following calculations to make comparisons later on to momentums which do involve relativistic velocities.

An important point about this stage of the process is that the m_1 blocks are still locked in place relative to m_2 and are sliding across the frictionless platform surface. The springs are still compressed. Due to *(103)* and the Law of Conservation of Momentum, the velocity of m_2 is equal in magnitude and opposite in direction to the velocity of m_3. The magnitude of each velocity is B_y. The momentum of the machine is:

$$P_1 = \frac{m_1 c B_y}{(1 - B_y^2)^{\frac{1}{2}}} = \text{momentum of each } m_1 \text{ block}$$

$$P_2 = \frac{m_2 c B_y}{(1 - B_y^2)^{\frac{1}{2}}} = \text{momentum of } m_2$$

$$P_3 = -\frac{m_3 c B_y}{(1 - B_y^2)^{\frac{1}{2}}} = \text{momentum of } m_3$$

$$P_s = -\frac{m_s c B_y}{(1 - B_y^2)^{\frac{1}{2}}} = \text{momentum of each spring's potential energy}$$

$$P_m = 2P_1 + P_2 + P_3 + 2P_s = \text{momentum of entire machine}$$

$$P_m = (2m_1 + m_2 - m_3 - 2m_s)\left[\frac{cB_y}{(1 - B_y^2)^{\frac{1}{2}}}\right] = 0 \qquad (104)$$

The momentum of the machine as a whole is still zero, which is in agreement with the initial system momentum. The machine is divided into two parts, and each part is moving with speed B_y. At this point, the grabbing mechanism holding the m_1 blocks in place is released. Even though there is a relative movement between the m_1 blocks and the platform, the construction of the devices will allow the springs to give the m_1 blocks a velocity B_x across the tracks (B_x is a speed measured relative to a reference frame stationary on m_2). After the devices have given velocities to the m_1 blocks, the situation is similar to that of Figure 19. The potential energy stored in the springs is converted into kinetic energy of the blocks. Intuitively, this discussion will be more easily understood if one assumes the velocities of the blocks are very high (i.e., near the speed of light). This will mean the mass property of the kinetic energy of the blocks or the potential energy of the compressed spring is significant when compared to m_2 or m_3. During the period of time when the blocks are not touching any platforms, as in Figure 19, the momentum of the m_1 blocks is:

$$P'_1 = \frac{m_1 c B_y}{(1 - B_x^2 - B_y^2 + B_x^2 B_y^2)^{\frac{1}{2}}} \quad \text{y direction momentum only} \quad (105)$$

$$P'_1 > P_1$$

The momentum P'_1 is obtained using the velocity of the block from the discussion of equations *(63)* through *(66)* That discussion does not use the concept of mass property, but uses the relativistic velocity transformations. These two discussions are similar in the momentums of the m_1 blocks. The momentum of the m_1 blocks can also be obtained using mass property concepts.

$$\frac{m_1 c^2}{(1 - B_x^2)^{\frac{1}{2}}} - m_1 c^2 = E_k = \text{kinetic energy of blocks as seen from a reference frame attached to } m_2$$

$$\frac{m_1}{(1 - B_x^2)^{\frac{1}{2}}} - m_1 = \frac{E_k}{c^2} = m_k = \text{mass property of blocks' kinetic energy, } m_2 \text{ reference frame}$$

$m_t = m_k + m_1$ = total mass of each block, m_2 reference frame

$$P'_1 = \frac{m_t c B_y}{(1 - B_y^2)^{\frac{1}{2}}}$$ = y direction momentum of total block mass in x-y reference frame

$$P'_1 = \frac{\dfrac{m_1}{(1 - B_x^2)^{\frac{1}{2}}} c B_y}{(1 - B_y^2)^{\frac{1}{2}}} = \frac{m_1 c B_y}{(1 - B_x^2 - B_y^2 + B_x^2 B_y^2)^{\frac{1}{2}}} \qquad (106)$$

The momentum of the entire machine can now be found. The momentums of m_2 and m_3 have not changed. The momentum of the springs' potential energy is now zero, since there is no spring potential energy in the situation of Figure 19.

$$P'_m = 2P'_1 + P_2 + P_3$$

$$P'_m = \left[2(m_1 + m_k) + m_2 - m_3\right]\left[\frac{cB_y}{(1 - B_y^2)^{\frac{1}{2}}}\right]$$

$E_s = m_s c^2$ potential energy spring had at maximum deflection

$$E_s = E_k$$

$$m_s = m_k$$

$$P'_m = (2m_1 + m_2 - m_3 + 2m_s)\left[\frac{cB_y}{(1 - B_y^2)^{\frac{1}{2}}}\right] \qquad (107)$$

Equation *(107)* uses the fact that $E_s = E_K$. The fact that the springs' potential energy is converted into the blocks' kinetic energy is not complicated by the fact that the blocks and springs have a relative motion with m_2 and m_3. E_s is the energy of the spring relative to a frame attached to m_3, and E_K is the energy of the block relative to a frame attached to m_2. The device of Figure 18 is arranged so that the relative motion between these two reference frames is not important. The device performs as if the spring were transferring its energy directly to the block, both being in the same reference frame.

A comparison of *(107)* and *(104)* reveals that the momentums of the two equations are not equal. P_m is equal to zero, and P'_m is clearly something greater than zero. There is no error. The machine momentum in this process is NOT conserved, even considering that there are no forces exterior to the system acting on the system. If the mechanism within m_3 that controls the relative m_2-m_3 motion brings the relative m_2-m_3 velocity to zero and locks m_2 and m_3 together, the entire machine will have a momentum in the positive y direction.

$$2m_1 + m_2 = m_3 + 2m_s$$

$$P'_m = \left[(m_3 + 2m_s) + (-m_3 + 2m_s)\right]\left[\frac{cB_y}{(1 - B_y^2)^{\frac{1}{2}}}\right]$$

$$P'_m = \frac{4m_s\, cB_y}{(1 - B_y^2)^{\frac{1}{2}}} \qquad (108)$$

Equation *(108)* applies only if m_2 and m_3 are locked together (no relative motion) before the m_1 blocks cross the tracks and compress the springs on the opposite sides. With m_2 and m_3 locked together before m_1 runs into the opposite platforms, the machine momentum given by *(108)* will be constant as the m_1 blocks run into the platforms and compress the springs. At maximum spring compression, the grabbing mechanism is activated to lock the m_1 blocks in position relative to m_2. The springs are at maximum potential energies again. The entire machine is now in the situation illustrated in Figure 21.

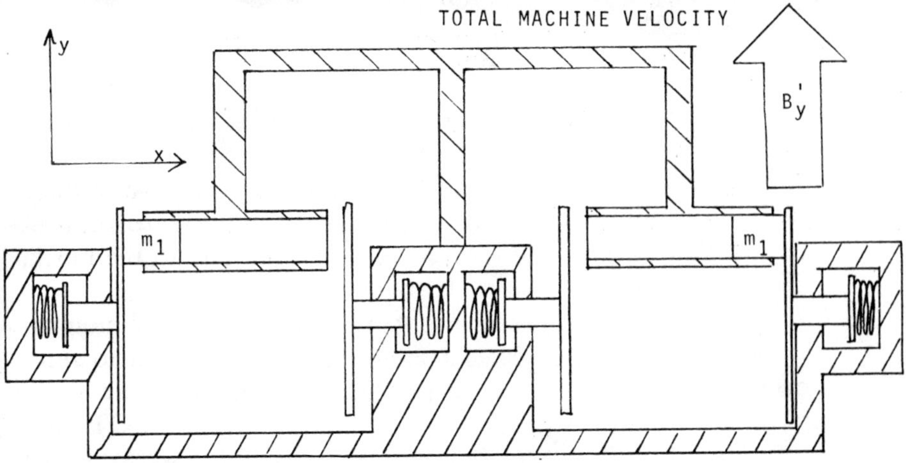

Figure 21

The velocity of the machine, B'_y, can be found by equating P'_m to the total machine momentum.

$$\frac{4m_s c B_y}{(1 - B_y^2)^{\frac{1}{2}}} = \frac{(2m_1 + m_2 + m_3 + 2m_s)cB'_y}{(1 - B'^2_y)^{\frac{1}{2}}} \tag{109}$$

With the m_1 blocks locked in position against the platforms, the relative positions of m_2 and m_3 are changed (by the mechanism within m_3) to the situation illustrated in Figure 22. During the transition from Figure 21 to Figure 22, the m_1 blocks slide down the platforms while holding the springs in a position of maximum compression. The momentum of the machine as a whole is constant during this process (due to the Law of Conservation of Momentum), though the individual momentums of the parts may change slightly as the movement from the situation of Figure 21 to Figure 22 is taking place. When the machine arrives at the situation of Figure 22, the mechanism within m_3 brings the relative m_2-m_3 velocity to zero, and locks the two together. This situation is equivalent to the situation of Figure 20. That is, it does not matter if the blocks are at their inside positions or outside positions (as shown in Figure 22). The process described by *(103)* to *(109)* is the same whether the blocks start at the inside positions and end up in the outside positions, or start in the outside positions and go through the process, ending up at the inside positions. Because Figure 22 can be taken to be the starting point for a repetition of the process of Figure 20, Figure 19, Figure 21, and Figure 22, the progression of events described by those figures makes up a cycle. This cycle produces the momentum of *(109)* or *(108)*. The velocity of the machine may be increased to any desired value by repeating the cycle as many times as is required. This cycle will be referred to as the momentum cycle.

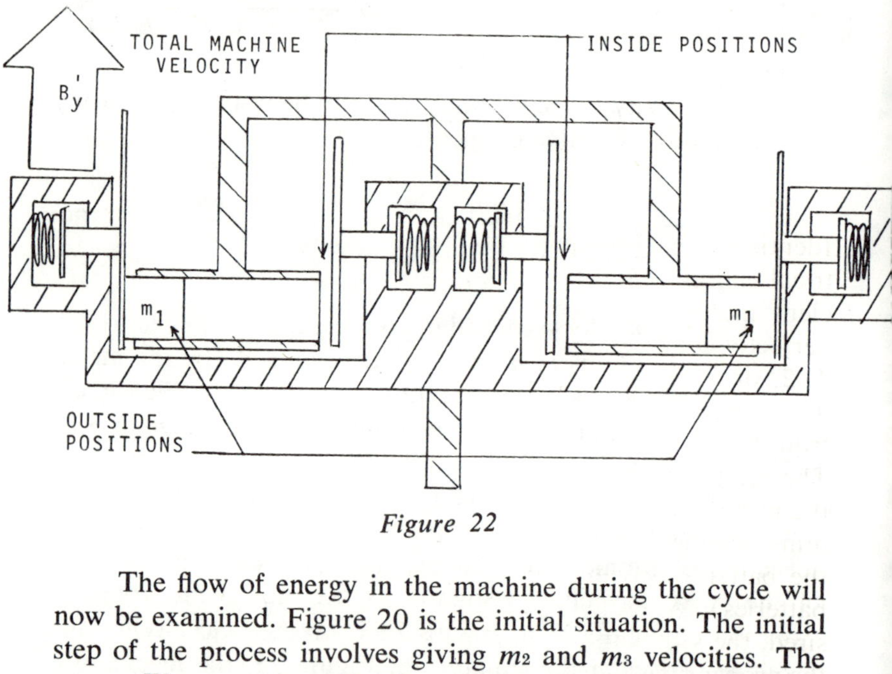

Figure 22

The flow of energy in the machine during the cycle will now be examined. Figure 20 is the initial situation. The initial step of the process involves giving m_2 and m_3 velocities. The energy W necessary to give the velocities is not critical in showing the machine momentums, but will be significant in this discussion. It is also necessary to have half of W coming from m_2 and half from m_3 in this more exact analysis. Before W is expended, the total mass of m_2 is $m_2 + W/2c^2$ and of m_3 is $m_3 + W/2c^2$. The mechanism needed to convert energy W into kinetic energy could be as illustrated in Figure 23, for example.

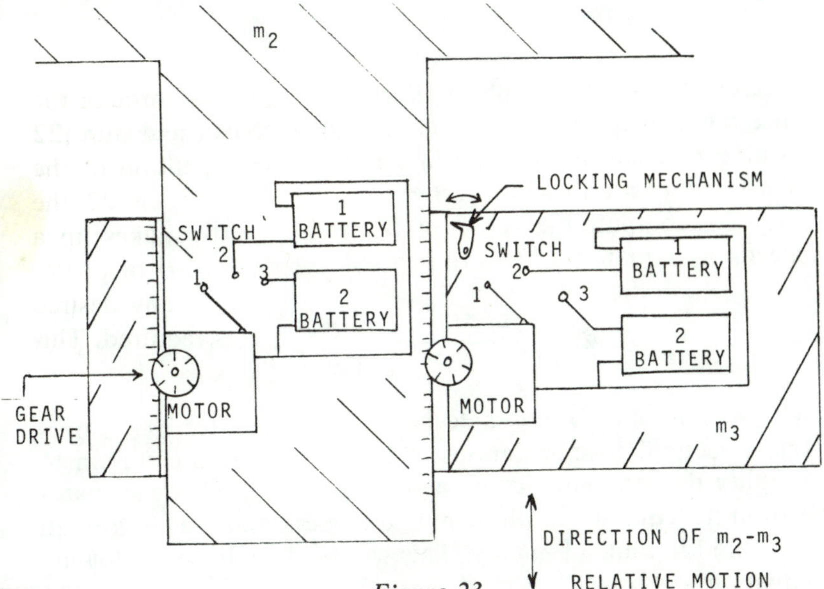

Figure 23

Identical apparatus on m_2 and m_3 is used to convert $W/2$ electrical energy to kinetic energy in each. Each has a 100%-efficient motor and two 100%-efficient batteries connected by perfectly conducting wires. A switch has an option of position *1*, *2*, or *3*. In position *2*, battery *1* gives motor torque in one direction, and in position *3*, battery *2* (with polarity reversed from battery *1*) gives a motor torque in the opposite direction. The relationship between the direction of torque generated by the motor and the direction of rotation of the motor will determine whether the motor acts as a motor (taking energy from the batteries) or a generator (putting energy back into the batteries). When the relative motion between m_2 and m_3 is desired, the switches are thrown simultaneously to position *2* for the proper time interval and energy W is taken from the batteries and converted completely into the required kinetic energies via the motors. The switches are returned to position *1* to take the batteries out of the circuits afterwards. The pole section of m_2 which passes through m_3 contains the motors and batteries for m_2. Switching to one battery or the other will give the relative m_2–m_3 velocity one direction or the other. Switching to a battery different from the one that initiates the motion will tend to reduce the velocity. In position *1*, the system coasts, and with the locking mechanism engaged into the gear teeth, the positions of m_2 and m_3 are locked together.

To compute the value of W needed to give m_2 and m_3 speeds B_y, equation *(37)* is to be used.

$$\triangle\left[\frac{mc^2B}{(1 - B^2)^{\frac{1}{2}}} + \frac{mc^2}{(1 - B^2)^{\frac{1}{2}}}\right] = 0 \qquad (37)$$

The version of *(37)* presented here is simplified for use in only one direction. In calculations of this nature, it is common to simplify the mathematics by assuming $c = 1$. This eliminates c from the equations. Momenta, energies, and masses are all measured in units of energy. This will be done in the following equations.

The situation of Figure 20 is represented in Figure 24. Figure 24a is the initial state of the system with all of the masses together with no velocities. In Figure 24b, m_2 and m_1 have positive B_y velocities, and m_3 and m_s have negative B_y velocities. Work W has been expended. In terms of *(37)*:

$$2m_1 + m_2 = m_3 + 2m_s \qquad (103)$$

final (momentum + energy)
− initial (momentum + energy) = 0

$$(2m_1 + m_2)\frac{B_y}{(1 - B_y^2)^{\frac{1}{2}}} - (m_3 + 2m_s)\frac{B_y}{(1 - B_y^2)^{\frac{1}{2}}}$$

$$+ (2m_1 + m_2 + m_3 + 2m_s)\frac{1}{(1 - B_y^2)^{\frac{1}{2}}}$$

$$- 0 - (2m_1 + m_2 + m_3 + 2m_s + W) = 0$$

$$W = (2m_1 + m_2 + m_3 + 2m_s)\left[\frac{1}{(1 - B_y^2)^{\frac{1}{2}}} - 1\right] \qquad (110)$$

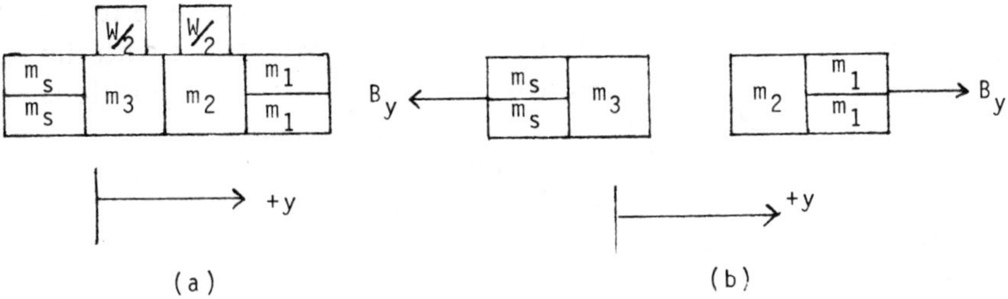

Figure 24

The process proceeds from Figure 20, where m_2 and m_3 have speeds B_y, to the situation of Figure 19, where m_2 and m_3 also have speeds B_y. In Figure 19, the springs' potential energy has been converted into the kinetic energy of the m_1 blocks. The mass property of the springs' potential energy is no longer connected with the movements of m_3, but has become the kinetic energy mass property, m_K, of the m_1 blocks and connected with the movements of m_2. This situation is illustrated in simplified form by Figure 25a. The process proceeds to a situation where there is no relative m_2-m_3 motion, but there is a velocity to the entire machine of B'_y, which is illustrated in Figure 25b. The reference frame that moves with velocity B'_y is commonly called the center of momentum reference frame. In chapter two, the Law of Conservation of Momentum was written immediately after inspection of the system, if the momentum of energy were included in the system momentum. Equation *(111)* is just this procedure for the process of Figure 25 as observed from the center of momentum reference frame. Equation *(111)* can also be found by using *(37)* in the primed reference frame and remembering the total machine momentum is zero in the center of momentum frame.

$$(2m_1 + m_2 + m_3 + 2m_s + W')\frac{B'_y}{(1 - B'^2_y)^{\frac{1}{2}}}$$

$$= (2m_1 + m_2 - m_3 + 2m_s)\frac{B_y}{(1 - B^2_y)^{\frac{1}{2}}} \qquad (111)$$

The work gains and losses of m_3, W' and W, are to be included in the energy calculations of the system (they were negligible in the momentum calculations). As the relative m_2-m_3 velocity is changed to zero, due to the mechanism of Figure 23, m_3 and m_2 will gain energy W' due to the change in kinetic energies of the machine parts. Energy W' can be determined from *(111)* and the following equation, which is equation *(37)* for the process of Figure 25 in the stationary reference frame.

$$0 = (2m_1 + m_2 + m_3 + 2m_s + W')\frac{B'_y}{(1 - B'^2_y)^{\frac{1}{2}}}$$

$$+ (2m_1 + m_2 + m_3 + 2m_s + W')\frac{1}{(1 - B'^2_y)^{\frac{1}{2}}}$$

$$- (2m_1 + m_2 + 2m_s)\frac{B_y}{(1 - B^2_y)^{\frac{1}{2}}} - \frac{m_3(-B_y)}{(1 - B^2_y)^{\frac{1}{2}}}$$

$$- (2m_1 + m_2 + m_3 + 2m_s)\frac{1}{(1 - B^2_y)^{\frac{1}{2}}}$$

$$W' = (2m_1 + m_2 + m_3 + 2m_s)\left[\frac{(1 - B'^2_y)^{\frac{1}{2}}}{(1 - B^2_y)^{\frac{1}{2}}} - 1\right] \qquad (112)$$

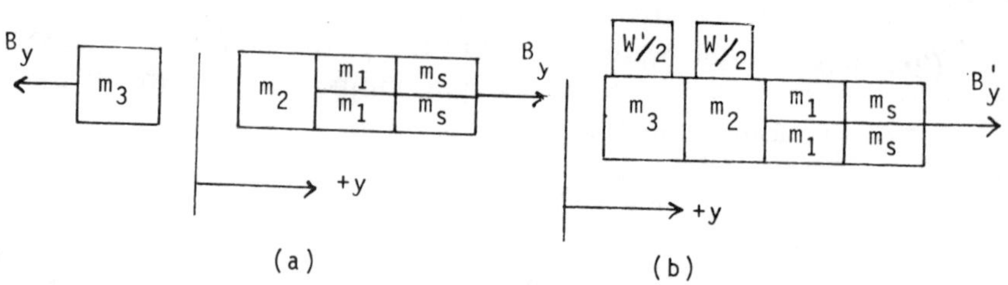

Figure 25

The energy and kinetic energy of the total machine in the situation of Figure 25b are:

$$\text{Energy} = E = (2m_1 + m_2 + m_3 + 2m_s + W')\frac{1}{(1 - B'^2_y)^{\frac{1}{2}}}$$

$$\begin{aligned}\text{Kinetic Energy} &= KE \\ &= (2m_1 + m_2 + m_3 + 2m_s + W')\left[\frac{1}{(1 - B'^2_y)^{\frac{1}{2}}} - 1\right] \quad (113)\end{aligned}$$

At this point, the process proceeds to the situation of Figure 21. The entire machine has velocity B'_y, but the mass property of the blocks' kinetic energy has been changed back into the potential energy mass property of the compressed springs of m_3. The energy of the machine of Figure 21 is also given by *(113)*. The last phase of the process is given by Figure 22, which is essentially the same situation as the initial situation of Figure 20 except that the machine has velocity B'_y. There is no relative m_2–m_3 velocity. In going from the situation of Figure 21 to that of Figure 22, no net energy is expended by the mechanism of m_3. The energy of the machine in the situation of Figure 22 is also given by *(113)*.

In going through one complete cycle, the machine expends energy W, gets back energy W', and has kinetic energy KE as a result. If the energy of this process is conserved throughout this cycle, then W less the kinetic energy KE should be equal to W'.

$$W - KE = W'$$

$$(2m_1 + m_2 + m_3 + 2m_s)\left\{\left[\frac{1}{(1 - B^2_y)^{\frac{1}{2}}} - 1\right] - \left[\frac{1}{(1 - B'^2_y)^{\frac{1}{2}}} - 1\right]\right\}$$

$$- W'\left[\frac{1}{(1 - B'^2_y)^{\frac{1}{2}}} - 1\right] = W'$$

$$= (2m_1 + m_2 + m_3 + 2m_s)\left[\frac{(1 - B'^2_y)^{\frac{1}{2}}}{(1 - B^2_y)^{\frac{1}{2}}} - 1\right]$$

$$(2m_1 + m_2 + m_3 + 2m_s)\left[\frac{1}{(1 - B_y^2)^{\frac{1}{2}}} - \frac{1}{(1 - B'^2_y)^{\frac{1}{2}}}\right]$$

$$= (2m_1 + m_2 + m_3 + 2m_s)\left[\frac{1}{(1 - B_y^2)^{\frac{1}{2}}} - \frac{1}{(1 - B'^2_y)^{\frac{1}{2}}}\right] \quad (114)$$

Although the momentum of the machine is not maintained during this process, the energy of the process is conserved.

The last step in the analysis of the momentum cycle is to see if the overall machine behavior follows the law of *(37)*. For the entire process, this would involve the change in energy and momentum of the machine with the situation of Figure 20 as the initial situation and the situation of Figure 22 as the final situation.

$$(W' + 2m_1 + m_2 + m_3 + 2m_s)\frac{B'_y}{(1 - B'^2_y)^{\frac{1}{2}}}$$

$$+ (2m_1 + m_2 + m_3 + 2m_s + W')\frac{1}{(1 - B'^2_y)^{\frac{1}{2}}}$$

$$- 0 - (2m_1 + m_2 + m_3 + 2m_s + W) = 0$$

$$B'_y = 0 \quad (115)$$

The overall energy-momentum relationship of the momentum cycle does not conform to *(37)* unless B'_y is zero, which is not the case for this cycle.

The discussion will now concern another variation of the machine of Figure 9b and Figures 19, 20, 21, and 22. In chapter three, the gravitational interactions of energy with matter and energy were discussed. The machine developed in the following text relies on this force for its operation. As before, this machine will be built gradually with a discussion concerning the operation of its parts. First, the device of Figure 18 will be used, and all discussion surrounding Figure 18 applies in this case. The m_1 blocks, previously discussed, will be used again in the same manner. The framework m_2 of the machine of Figure 19 is to be used; the m_1 blocks and devices require a framework of that type for their operation. The mechanism to grab the m_1 blocks at each end of their tracks is also carried over. The final part of the new machine is a mass similar to m_3 of Figure 19, with the devices located as they are in Figure 19. The mechanism governing the relative m_2–m_3 movement is still present, as described previously. The machine of this discussion is, in fact, identical in its parts to the previous machine. Also, the steps of the process of the previous discussion are almost the same. The cycle described below is not the same cycle as the previous discussion.

In the following discussion, the first concern will be with the flow of energy within the machine. The initial situation of the machine is illustrated in Figure 26. The m_1 blocks are held

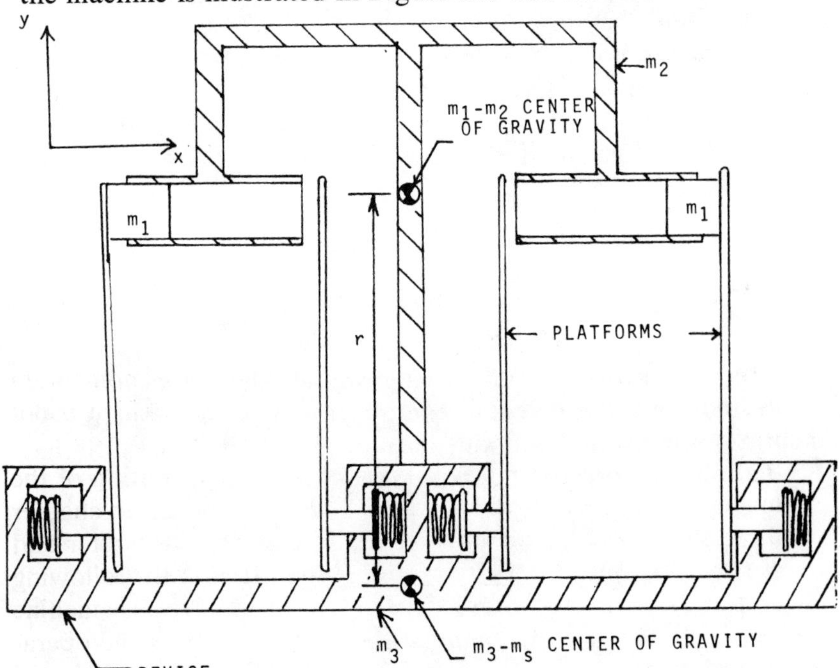

Figure 26

at their extreme positions in the tracks by the grabbing mechanism. The springs in the two devices are at their maximum compression. As before, m_s is the mass property of a compressed spring's potential energy and m_K is the mass property of the m_1 block's kinetic energy. The m_2-m_3 positioning is as shown, with m_2 as far from m_3 as the limits of the machine allow. This position is held by the locking mechanism mentioned in the description of Figure 23. There is a gravitational attraction between the springs' potential energy and mass m_3, and the m_1 masses and mass m_2. In Figure 26, the devices are drawn in a different location on m_3 to denote that the springs' potential energy will be involved with the gravitational interactions of the system. The gravitational forces in the y direction are the only forces to be considered here. The machine of Figure 26 is expanded in the y direction to emphasize this. The gravitational forces are assumed acting on the center of gravity of the m_1-m_2 system and the m_s–m_3 system, as if all of the mass of each machine part is concentrated at a single point. If r, of Figure 26, is large enough, this assumption is accurate.

The gravitational attraction between the two parts of the machine will put the following force on the positioning mechanism.

$$F_1 = \frac{G(2m_1 + m_2)(m_3 + 2m_s)}{r^2}$$

$$F_1 = \frac{G(2m_1m_3 + m_2m_3 + 2m_sm_2 + 4m_sm_1)}{r^2} \qquad (116)$$

At this point the grabbing mechanism holding the m_1 blocks is released, and the potential energy stored in the compressed springs is converted into kinetic energy of the m_1 blocks. In this situation, the force on the positioning mechanism in m_3 is:

$$m_s = m_k$$

$$F_2 = \frac{G(2m_1 + m_2 + 2m_s)(m_3)}{r^2}$$

$$F_2 = \frac{G(2m_1m_3 + m_2m_3 + 2m_sm_3)}{r^2} \qquad (117)$$

While the m_1 blocks are traveling across the tracks, the positioning mechanism decreases the distance between m_2 and m_3 to the minimum allowable within the limits of the machine. To simplify the mathematics, it will be assumed that m_3 is the earth, with a mass much larger than any of the other masses involved. This will make kinetic energy changes of the machine as a whole negligible, as the mass m_3 can be considered stationary at all times. In addition, the work added to m_3 in any of the process steps will have a mass property that is negligible when compared to the total mass of m_3. Assume the positioning mechanism inside m_3 is similar to that of Figure 23, except that only the m_3 half will be involved. With m_3 as large as the earth's mass, F_2 will be larger than F_1.

The work put into the positioning mechanism (i.e., the energy gain of the battery in Figure 23 when the motor is acting as a generator) will be determined as the relative m_2-m_3 position goes from its maximum to the minimum while the m_1 blocks are traveling across the tracks. It will be assumed the relative m_2-m_3 position goes through its complete travel from the maximum to the minimum distance before the m_1 blocks travel completely across the track. If R_1 is the maximum distance between m_2 and m_3, and R_2 is the minimum, the energy W_1 added to m_3 as the relative m_2–m_3 position goes from R_1 to R_2 is:

$$W_1 = -\int_{R_1}^{R_2} F_2 dr$$

$$W_1 = -\int_{R_1}^{R_2} \frac{G(2m_1 + m_2 + 2m_s)m_3}{r^2} dr$$

$$W_1 = G(2m_1 + m_2 + 2m_s)\, m_3 \left(\frac{1}{R_2} - \frac{1}{R_1}\right) \tag{118}$$

At this point, the relative m_2-m_3 velocity is brought to zero and the m_1 blocks run into the platforms at the end of the

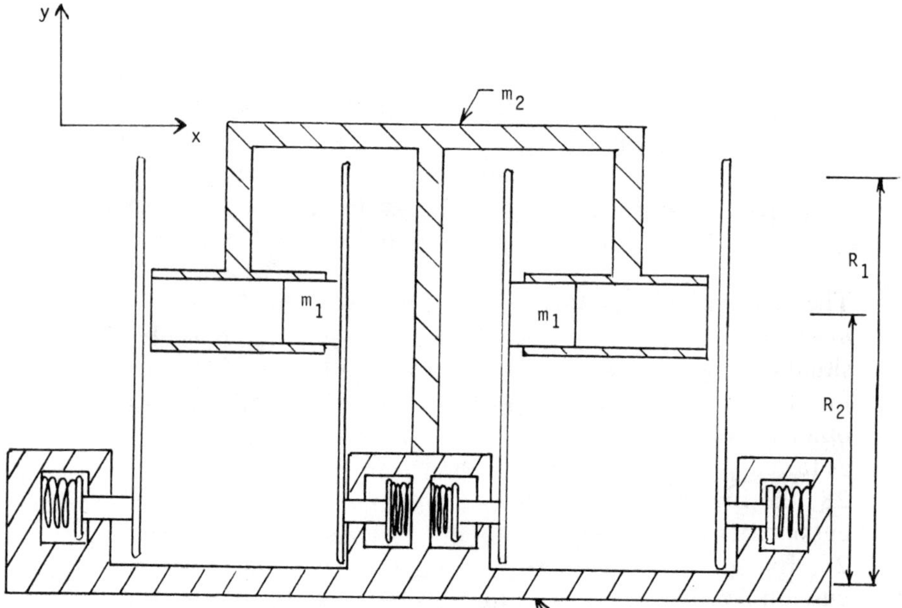

Figure 27

tracks, converting the kinetic energy of the m_1 blocks into potential energy of the compressed springs in the devices. When the compression of the springs is at its maximum, the grabbing mechanism in m_2 grabs the blocks and holds them. The configuration of the machine is illustrated in Figure 27. The gravitational forces within the machine generate a total force on the positioning mechanism given by *(116)* with $r = R_2$. With the m_1 blocks locked in position against the platforms by the grabbing mechanism, the positioning mechanism moves the relative m_2-m_3 position from R_2 back to R_1. The m_1 blocks will slide along the frictionless platform surfaces, maintaining the two springs involved at maximum compression. The work expended by the positioning mechanism as the relative m_2-m_3 position goes from R_2 to R_1 is:

$$W_2 = -\int_{R_2}^{R_1} F_1 dr$$

$$W_2 = -\int_{R_2}^{R_1} \frac{G(2m_1 + m_2)(m_3 + 2m_s)}{r^2}\, dr$$

$$W_2 = G(2m_1 + m_2)(m_3 + 2m_s)\left(\frac{1}{R_1} - \frac{1}{R_2}\right) \qquad (119)$$

The relative m_2-m_3 velocity is brought to zero as the relative m_2-m_3 positon reaches R_1. This is illustrated in Figure 28. This situation is the same as the initial situation of Figure 26. The fact that the m_1 blocks are at one end of the tracks versus the other is not a significant difference in the state of the machine as far as the analysis is concerned. From the situation of Figure 28, the machine is in position to repeat the steps of the discussion of *(116)* through *(119)*. Thus the process given by Figure 26, Figure 27, and Figure 28, together with the related discussion, constitute a complete cycle. During the cycle, the positioning mechanism of m_3 has first received work W_1 as

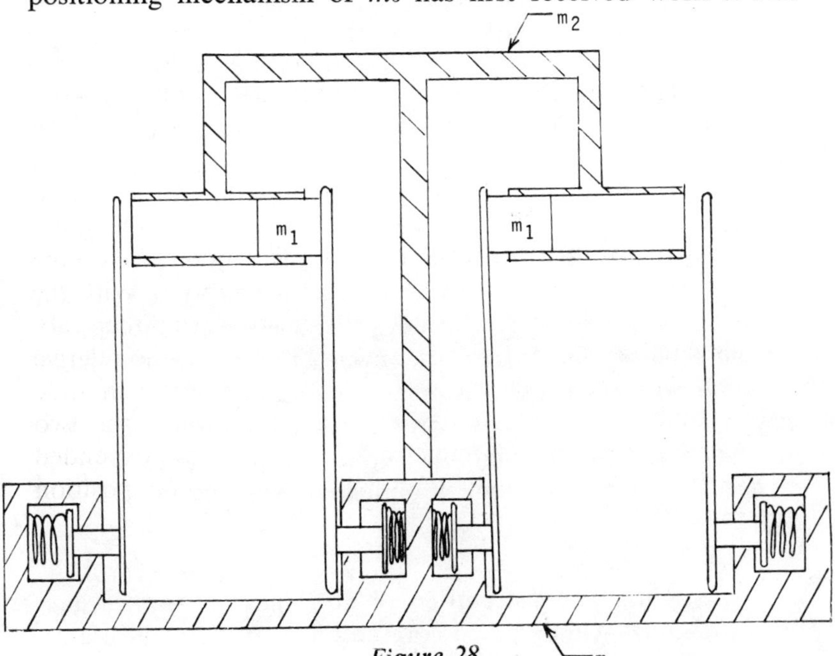

Figure 28

gravity pulls m_2 and m_3 closer together, and then has work W_2 taken away, as it restores m_2 and m_3 to their original positions. The net work input to the positioning mechanism is $W_1 + W_2$:

$$W_1 + W_2 = G(2m_1m_3 + 2m_sm_3 + m_2m_3)\left(\frac{1}{R_2} - \frac{1}{R_1}\right)$$
$$- G(2m_1m_3 + m_2m_3 + 2m_sm_2 + 4m_sm_1)\left(\frac{1}{R_2} - \frac{1}{R_1}\right)$$

$$W_1 + W_2 = 2Gm_s\,(m_3 - m_2 - 2m_1)\left(\frac{1}{R_2} - \frac{1}{R_1}\right) \qquad (120)$$

The net work input to the positioning mechanism will be zero if $m_3 = m_2 + 2m_1$. That is not a condition on the structure of the machine, however, but merely one of the possibilities for those mass values. There is no reason why m_3 can't be different from $2m_1 + m_2$, and this discussion has stipulated that m_3 is the earth's mass ($m_3 \gg 2m_1 + m_2$). The machine has undergone a true cycle (final and initial states are identical) but has gained $W_1 + W_2$ amount of energy without any energy being added from outside. This is in conflict with the Law of Conservation of Energy as presented. This cycle will be called the energy cycle. The proper variation of parameters and cycle design will produce a cycle that either produces or destroys energy. Apparently any amount of energy may be produced or destroyed by repeating the cycle for the machine of the proper design. A comparison of the energy-producing cycle and momentum-producing cycle reveals that energy and momentum can both be produced from one cycle. The proper parameter selection will determine the relative values of energy or momentum production. A properly constructed cycle will result in a machine that will continue to produce the amount of energy needed to produce momentum increases indefinitely.

$$W_1 + W_2 \geq W - W' \qquad (121)$$

In summary, constructions of machines and cycles have been presented which are in contradiction to the principles of the Law of Conservation of Energy and the Law of Conservation of Momentum. These two laws have been strictly observed at every cycle phase, but the overall machine performance has resulted in contradictions to the laws.

Chapter Five

PHILOSOPHY

Chapter one is significant because it stresses the difference between a measured quantity and an actual quantity. When a measurement does not reflect accurately the actual amount being measured, the notion of equality can no longer be defined with confidence, and there is an undermining of the mathematical system. Length contraction is the principle used. The discussion may not be clear if one considers only moving straight lines of a constant velocity because the conflicts involved do not produce immediate disasters. One merely reassures oneself by saying "moving objects shorten in the direction of motion" and the crisis is over.

The spinning disc example presents an immediate contradiction. The circumference must have different values at the same time depending upon the measurement method. Both circumference values are correct in the context of the measurement methods. The rocketship example also presents an immediate conflict because the speed of light is exceeded without much difficulty at all. These are not trivial examples. They reflect fundamental changes in the structure of the universe between two moving reference frames, and the inability of a mathematical idea (equals), which is unquestionable in a stationary reference frame, to be directly applicable in that circumstance. When one realizes the measurement may not accurately reflect the actual quantity being measured, the situations are seen to present no real conflicts, only apparent ones. The spinning disc was first used by Einstein in a paper entitled "The Foundation of the General Theory of Relativity" which appears in the book *The Principle of Relativity* by Dover Publications, Inc. Einstein did not draw the same conclusions from the example as were drawn here.

In chapter two, the momentum of energy is discussed. That energy has momentum is a concept presented previously in relativity theory, though the method used in chapter two is unique. The concept of mass property is introduced, primarily as an aid in the presentation of the mass of energy. The concept of total mass is also helpful in understanding of the mass-like characteristics of energy, and provides some new viewpoints for studying these characteristics. One important point that should be recognized is that all of the discussion of chapter two is based on the fact that matter can be converted into heat energy. This fact has been documented in many experiments. The tools developed in chapter two are also of great importance in the discussion of chapter four.

Chapter three presents the relationships between gravity and energy. Again, this is not a new concept, although the method used to present it is. The concept of mass property is of use in this chapter, too. As in chapter two, the discussion of chapter three relies primarily on the experimentally verified fact that matter may be changed into heat energy. This chapter is also significant in developing tools for use in chapter four.

Chapter four presents a machine that changes energy into momentum, and a machine that produces energy. The machines of this chapter rely heavily on the discussions in chapters two and three because those chapters have such a close relationship with actual experimental results (the changing of matter to heat through nuclear fission). There is, unfortunately, no direct experimental verification of the cycles presented.

The machine of the momentum cycle relies on the device of Figure 18. That is the key to the machine's operation because it allows the machine to change the velocity of the mass property of energy without penalty (spring-energy mass property has a velocity in one direction, and is converted into kinetic energy which has a velocity in the other direction). This results in a momentum change without an applied force.

The machine of the energy cycle also relies on the device of Figure 18. In this case, it allows the transfer of mass property from one gravitational potential to another without the expenditure of work (again as the springs' potential energy mass property changes to the moving blocks' kinetic energy mass property, each mass property being located at a different gravitational potential). The result is that this energy can produce work as it returns to its original potential.

These chapters are all independently important, but the hypothesis will now be that they are also important as a group. Chapters two, three, and four are obviously related and interdependent, but the hypothesis is that chapter one has some important points in common with the other three. Chapter one deals primarily with length and chapters two, three, and four deal primarily with mass and energy. The first area they have in common is that all of the chapters rely on the Theory of Relativity for their existence. In relativity theory, another link has already been established between dimension and mass through the Law of Conservation of Momentum. That link is the requirement that the Law of Conservation of Momentum be invariant under the Lorentz transformations [equations *(34)*, *(35)*, and *(36)*].

Using that idea as some substantiation to the above hypothesis, another common ground between the chapters is sought. There are two main concepts here. One is that measured and actual dimensions are not always equal, and the other is that momentum and energy are not always conserved. It is difficult, at first, to pinpoint what these two concepts have in common, but after some consideration one may reach the conclusion that in both cases something is "wrong," or, at least, not as expected. Things aren't working out as "common sense" would have them. (That is one of the main difficulties a beginner in relativity theory often encounters.) The same line has many lengths, the speed of light appears to be exceeded, momentum is created (or destroyed), and energy is created (or destroyed). These are not the results that one's training has taught one to expect.

The key relationship between these ideas deals with the most fundamental assumption of scientific study. A brief look into the history of mathematics will be offered as part of this explanation. Mathematics is a convenient notation that man has devised for arranging thoughts, storing information, and extending the basic abilities of his mind. But the main reason for the existence of mathematics is to describe the universe. If someone has one apple and a friend gives him another apple, he has two apples. That's an elementary example to be sure, but even the most elementary thoughts are difficult to express without mathematical principles.

Man finds himself in a universe, and he seeks to describe and further understand this universe, so he invents mathematics. Throughout its history, mathematics is refined and becomes much better at describing the universe. It allows man to express complex thoughts, to reason, to form abstractions and make predictions. Mathematics has become very powerful, and so in tune with the universe that subtly the assumption has been made that the universe behaves mathematically. This is not a bad assumption, since the overwhelming majority of experimental evidence has absolutely supported the assumption.

Recently, however, some information has been gathered which was not available when mathematics was being formulated. The information is based on experimental evidence and places some totally new conditions on the universe. "No object can go faster than the speed of light." "There is no absolute reference frame." Previously, the inability of one facet of mathematics (equals) to adapt to these assumptions has been explained. The assumption that the universe behaves mathematically in every situation is questionable under this circumstance.

"Momentum is always conserved." That is a statement that makes sense in the context of the mathematically based logic that has been a part of scientific study. If someone has four apples, and nobody gives him any apples, he can never have five apples. Four always equals four. Four never equals five, or three, or thirty-six. How is it possible that one can end up with more than he started with? It doesn't make sense in the range of experiences one has had up to this point it time.

Mathematics, as devised by man, has a very distinctive type of perfection and inflexibility. But mathematics is man's way of describing the universe, and the sophistication of mathematics at any point in time reflects, to some degree, the amount of knowledge that man has gained to that point. That does not necessarily imply that the universe must be perfect in the same way that mathematics is perfect. The creation of energy and momentum and discrepancies in lengths and velocities may not be ideas that are consistent with mathematical logic, but mathematics does not tell the universe how to operate, it just describes the universe to the best of man's abilities.

Some experiments have been devised that are not consistent with mathematically based logic. These examples involve creation, destruction, and other inconsistencies. Mathematics is a form of logic designed to be free of inconsistencies. These things cannot happen in a perfect mathematical universe. This is what chapter one has in common with chapters two, three, and four. The refined version of the hypothesis is that because ideas may be in conflict with mathematical thought, that, in itself, is not proof enough that the ideas cannot actually be consistent with the operation of the universe.